Springer Theses

Recognizing Outstanding Ph.D. Research

For further volumes:
http://www.springer.com/series/8790

Aims and Scope

The series "Springer Theses" brings together a selection of the very best Ph.D. theses from around the world and across the physical sciences. Nominated and endorsed by two recognized specialists, each published volume has been selected for its scientific excellence and the high impact of its contents for the pertinent field of research. For greater accessibility to non-specialists, the published versions include an extended introduction, as well as a foreword by the student's supervisor explaining the special relevance of the work for the field. As a whole, the series will provide a valuable resource both for newcomers to the research fields described, and for other scientists seeking detailed background information on special questions. Finally, it provides an accredited documentation of the valuable contributions made by today's younger generation of scientists.

Theses are accepted into the series by invited nomination only and must fulfill all of the following criteria

- They must be written in good English.
- The topic should fall within the confines of Chemistry, Physics and related interdisciplinary fields such as Materials, Nanoscience, Chemical Engineering, Complex Systems and Biophysics.
- The work reported in the thesis must represent a significant scientific advance.
- If the thesis includes previously published material, permission to reproduce this must be gained from the respective copyright holder.
- They must have been examined and passed during the 12 months prior to nomination.
- Each thesis should include a foreword by the supervisor outlining the significance of its content.
- The theses should have a clearly defined structure including an introduction accessible to scientists who are not expert in that particular field.

Evan Francis Keane

The Transient Radio Sky

Doctoral Thesis accepted by
The University of Manchester, UK

Author
Dr. Evan Francis Keane
School of Physics and Astronomy
Jodrell Bank Centre for Astrophysics
The University of Manchester
Alan Turing Building, Oxford Road
Manchester M13 9PL
UK
e-mail: Evan.Keane@gmail.com

Supervisor
Prof. Michael Kramer
Jodrell Bank Centre for Astrophysics
The University of Manchester
Alan Turing Building, Oxford Road
Manchester M13 9PL
UK
e-mail: Michael.Kramer@manchester.ac.uk

ISSN 2190-5053
ISBN 978-3-642-19626-3
DOI 10.1007/978-3-642-19627-0
Springer Heidelberg Dordrecht London New York

e-ISSN 2190-5061
e-ISBN 978-3-642-19627-0

Cover design: eStudio Calamar, Berlin/Figueres

Printed on acid-free paper

Springer is part of Springer Science+Business Media (www.springer.com)

Dedicated to Sharon, my one and only

Supervisor's Foreword

For millennia, humans have studied the sky and its phenomena. Often seemingly mysterious sightings were described and interpreted in certain ways. Today, we still do this, but now equipped with powerful telescopes that allow us to peek at the edge of the visible Universe. We now use the Cosmos as our gigantic laboratory to understand the birth and evolution of the Universe, to study and understand the nature of fundamental forces, and to find our place in this vast space. We are amazingly successful, being able to describe the almost complete history of the Universe, back in time until a very small fraction of a second before the Big Bang. If this were not dazzling enough already, we achieve this only by registering and detecting photons with our electronic eyes and devices installed on our modern telescopes. But, our description of this Universe would be incomplete, if we hadn't made a major step in our cosmic exploration: we do not only gather information in the tiny window of the electromagnetic spectrum that covers visible light, but we receive photons across the whole spectrum, from low-energy radio waves to high-energy gamma-rays. It is this combination of spectral information that really reveals the nature of the cosmic processes and objects.

Very often, objects may only be visible or easily detectable at particular frequencies, for instance at radio frequencies. Indeed, the objects known as "pulsars" are a superb example. Pulsars are the remnants of exploded massive stars, representing the dense, collapsed core of those, where gravity has squeezed protons and electrons into neutrons. Such "neutron stars" consist of the most extreme and densest matter in the observable Universe—observable because we can often detect them as pulsars. Neutron stars which appear as pulsars are highly magnetized, rotate swiftly around their axis in seconds or much less, and emit a narrow beam of radio emission along their magnetic poles. As the pulsars rotate, they act as cosmic lighthouses, sweeping a beam of radio light across the Universe and appearing as a pulsating radio source to an observer happening to be in the path of the sweeping beam.

Pulsars are very useful for a number of exciting experiments involving, for instance, gravitational physics, but we can only see and use them, if the beam is directed towards Earth during their rotation. Having studied them now for more than 40 years after their accidental discovery in Cambridge in 1967 by Jocelyn Bell and her Ph.D. supervisor Anthony Hewish, we have a pretty good estimate on how many active pulsars exist in the Milky Way—even if not seen—and how many neutron stars exist in general. At least we thought so…

Only in 2006 researchers at Jodrell Bank Observatory of the University of Manchester discovered a new type of neutron star that did not seem to fit the description of a radio pulsar. Rather than emitting regular and periodic pulses like pulsars, they only emitted short flashes of radio light, lasting only for fractions of milliseconds or so. The discovery was lucky and by chance: one had to look at the right moment of time at the right spot on the sky to detect these radio flashes, as the total amount of emitted radio light was usually less than a second per day. The lead investigator of this discovery, Maura McLaughlin, named these objects "Rotating Radio Transients" or RRATs as data soon suggested that these new sources were also rotating neutron stars, but where the emission is not steady or purely periodic but transient in nature. Initial estimates suggested that the number of RRATs may exceed the number of active pulsars by a significant factor, suddenly changing our previous understanding about the Galactic neutron star population. Further studies were clearly needed.

This thesis by Evan Keane is a direct and important continuation of these studies. While Maura with our team and others worked on many different aspects to establish the properties of RRATs, Evan's task was to quantify selection effects for determining the real number of RRATs (and hence, neutron stars in general!) and to explore their relationship to "ordinary" radio pulsars. His findings, described in this excellent piece of work, were surprising to us. They change our definition of RRATs, making clear that RRATs and pulsars are not really different but part of the same family. Furthermore, Evan established without doubt that there are indeed lots of them that await our discovery.

Based on Evan's work, it is now our task to find the "missing" RRATS and to confirm the important conclusions of this thesis. This will not be easy and will require new telescopes with much better sensitivity and sky coverage, but we will do it! And, of course, these telescopes will then again also find something unexpected, continuing our exciting journey through this truly wonderful Universe.

January 2011 Michael Kramer

Acknowledgments

I would like to thank my supervisor, Prof. Michael Kramer for giving me the opportunity to do an interesting research project in a fantastic group.

Thank you to Sharon, mo ghŕa, mo chara, for helping me in every aspect of my life, for giving me the strength to persevere when all I wanted to do was quit, run and hide, for picking me up when I was down, and for always being in arrivals when I returned from my travels.

I have benefited greatly from so many people in the last 3 years. When I started, the resident Jodrell postdocs, Aris Noutsos and Roy Smits, helped me a lot, answering my silly questions. Thanks to Sam Bates for talking all things pulsar, hydra, nerd and sport related with me. Thanks to all of those who organised basketball, in particular the Whitworth Park basketball team. I thank MarkPurver for organising (more) basketball, football, several Man. Utd. tickets and high-density car journeys to Jodrell Bank. Thanks to Cristobal Espinoza for all the helpful discussions about glitches and to Kuo Liu for incites into the mysteries of pulsar timing. Thanks to all my other office mates also—Tom Hassall, NeilYoung, Sotirios Sanidas, Phrudth "Math" Jaroenjittichai, Cees Bassa, Gemma Janssen, Rob Ferdman and Ralph Eatough.

Thanks to Ben Stappers for helping me with more things than I can remember. I would like to thank Andrew Lyne for the many meetings, phone calls and emails which helped to teach me about pulsar timing. Thanks also to Chris Jordan for all her help, and for providing me with data. A huge thanks to Anthony Holloway and Bob Dickson, for sorting out all my computer problems, and an extra thanks to Ant for organising FF1. I thank my pulsar friends scattered about the Earth—my sometimes housemate Maciej Serylak, Eduardo Rubio Herrera, Marta Burgay, the entire EPTA crew and John Sarkissian for going above and beyond the call of duty on more than one occasion to take observations on my behalf. Thanks to the many wise sages who I've had the pleasure of working with—Sir Francis Graham-Smith, whose knowledge on all things pulsar related simply blows me away, to Maura McLaughlin for discovering the spider-babies that are RRATs (!) and Duncan Lorimer to name but a few. Extra gratitude is due to the people who read early drafts of my thesis—Cess Bassa, Ben Stappers and Andrew Lyne.

Special thanks go to the person, or persons, who stole some of my data whilst it was in transit from China, and replaced the data discs with a Bangladeshi driving licence. I would like to thank the Church of the Flying Spaghetti Monster. Without the nurturing touch of their noodely appendages nothing would be possible.

Thank you to Luke, mo chara dhílis for hosting me in London, visiting me in Manchester and being a great friend. Thanks to everybody who came to one ofmy talks and to the anonymous referees out there whose comments increased my work load but helped improve the quality of my papers greatly.

I also happily acknowledge the financial support given to me as part of my Marie-Curie EST Fellowship with the FP6 Network "ESTRELA" under contract number MEST-CT-2005-19669. Thanks also to Neal Jackson for organising the ESTRELA Research Training Network. As well as funding my studies it has enabled me to travel to many observatories, universities and institutes throughout Europe, North America and Australia which I would never have been able to do otherwise. This has also meant that I have met a number of great people along the way—long live my fellow Estrelians!

Peace and Love

Contents

1	**Introduction: Radio Transients**	1
	1.1 Transient Radio Emission	2
	1.2 Transient Phase Space	4
	1.3 Sources of Transient Radio Emission	6
	1.4 Thesis Outline	11
	References	11
2	**Neutron Stars**	15
	2.1 The Life and Death of Stars	15
	2.2 Theory of Neutron Stars	18
	2.2.1 Neutron Stars on the Back of an Envelope	18
	2.2.2 Neutron Star Structure	19
	2.3 Observed Manifestations of Neutron Stars	27
	2.3.1 Pulsars	27
	2.3.2 RRATs	32
	References	37
3	**On the Birthrates of Galactic Neutron Stars**	41
	3.1 Introduction	41
	3.2 Different Manifestations of Neutron Stars	42
	3.3 Birthrates	45
	3.4 Too Many Neutron Stars?	50
	3.5 Discussion	52
	3.6 Conclusions	57
	References	58
4	**PMSingle: A Re-Analysis of the Parkes Multi-Beam Pulsar Survey in Search of RRATs**	61
	4.1 Observing Pulsars	61
	4.2 Single Pulse Searches	62

4.3 The Parkes Multi-Beam Pulsar Survey . . . 68
4.3.1 Why Re-process? . . . 68
4.3.2 The Zero-DM Filter . . . 70
4.4 PMSingle . . . 76
4.5 Detections and New Discoveries . . . 81
4.5.1 Detection of Known Sources . . . 83
4.5.2 New Discoveries: Repeating Sources . . . 86
4.5.3 New Discoveries: Non-Repeating Sources . . . 90
4.6 Discussion . . . 94
4.6.1 What is a RRAT? . . . 94
4.6.2 Distant Pulsars? . . . 95
4.7 Conclusions . . . 97
References . . . 98

5 Timing Observations of RRATs . . . 101
5.1 Pulsar Timing . . . 101
5.1.1 Integrated Profiles . . . 102
5.1.2 Single Pulses . . . 104
5.1.3 From Bits to BATs . . . 105
5.2 Timing at Jodrell Bank . . . 107
5.2.1 Pulsar Glitches . . . 109
5.2.2 J1819–1458 . . . 109
5.2.3 J1913+1333 . . . 115
5.2.4 The JBO DFB . . . 117
References . . . 121

6 Timing Solutions for Newly Discovered Sources . . . 123
6.1 PMSingle Timing Solutions . . . 123
6.1.1 Complete Timing Solutions . . . 123
6.1.2 Preliminary/Unsolved PMSingle Sources . . . 128
6.1.3 Original RRAT Timing Solutions . . . 129
6.1.4 New Discoveries . . . 130
6.1.5 An Aside: The Perils of EFAC . . . 131
6.2 Timing Status and Prospects . . . 132
6.2.1 Importance of Timing Solutions . . . 133
References . . . 135

7 X-ray and Optical Observations of RRATs . . . 137
7.1 J1840–1419: The Coolest Neutron Star Ever Known? . . . 137
7.2 Optical Observations of J1819−1458 . . . 140
7.2.1 Observations and Data Reduction . . . 141
7.2.2 Results . . . 143
7.2.3 Discussion . . . 146
References . . . 147

8 RRATs: An Overview 149
8.1 Recent Observations 149
8.2 Other Relevant Work and Ideas 153
References 155

9 Conclusions 157
9.1 What Do We Know Now? 157
9.2 When a Pulsar is a RRAT 158
9.2.1 Selection Effects 158
9.2.2 'Solutions' 161
9.2.3 Switching Magnetospheres? 162
9.2.4 The PMPS RRATs 164
9.3 Questions Answered 164
9.3.1 Facts Abouts RRATs 167
9.4 Future Work 168
References 169

Appendix A: Basic Equations of Radio Astronomy 171

Appendix B: Neutron Stars: Supplementary 175

Appendix C: Birthrates: Supplementary 177

Appendix D: PMSingle: Supplementary 179

Appendix E: Timing: Supplementary 181

The Author 185

Index 187

Chapter 1
Introduction: Radio Transients

What is actual is actual only for one time. And only for one place
T. S. Eliot

Short-timescale bursts, pulses, flares or flickering at radio frequencies signal extreme astrophysical environments. A pulse of width W, with a flux density S, at an observing frequency ν, which originates from a source at a distance D, has a brightness temperature of

$$T_B \geq 4 \times 10^{23}\ \mathrm{K} \left(\frac{SD^2}{\mathrm{Jy\ kpc}^2}\right)\left(\frac{\mathrm{GHz\ ms}}{\nu W}\right)^2. \tag{1.1}$$

The minimum T_B in this expression is obtained when the emitting region is the maximum size of a causally connected region, $cW = 300$ km $(W/1$ ms$)$. If the dynamical time $t_{\mathrm{dyn}} = \sqrt{1/G\rho}$ dictates the scale on which we see changes, where ρ is density and G is Newton's constant, then the millisecond radio sky consists mainly of neutron stars, which have $t_{\mathrm{dyn,NS}} \sim 0.1$ ms, whereas transient emission can be expected from white dwarfs on timescales of $t_{\mathrm{dyn,WD}} \sim 1$–$10$ s. Equation 1.1 is parameterised in units typical of Galactic bursts of millisecond duration. We can see that observations of the transient radio sky probe compact objects and coherent non-thermal emission processes.

At present there is a large number of transient radio sources known. This is despite the relatively unexplored 'transient phase space'. The known transient time-scales vary from as short as nano-seconds up to as long as months to years. This upper limit is quite arbitrary, and typically set by the time baseline of observations for a given object, whereas the lower limit is set by the physical processes occurring in known sources, as well as the technical constraint of being able to observe in the sub-nanosecond regime. In this chapter—we recap some of the relevant emission mechanisms, before discussing the transient parameter space, and to what degree it has been, and will be, investigated. We then describe known sources of transient radio emission as well as theoretical objects which have been proposed to exist. Finally, we briefly introduce neutron stars and RRATs, the major focus of the work presented in this thesis.

E. F. Keane, *The Transient Radio Sky*, Springer Theses,
DOI: 10.1007/978-3-642-19627-0_1,

1.1 Transient Radio Emission

Radio emission can be either thermal or non-thermal. The observed transient radio sources, which we outline below, all emit non-thermal radiation. This non-thermal radiation is itself commonly divided into either incoherent or coherent. If we consider electrons which radiate some fraction ζ of their energy, then the energy density of the radiation is $\zeta(\gamma - 1)nmc^2$, for number density n. The energy, $kT/V_{\rm coh}$, in a coherence volume[1] $V_{\rm coh}$, cannot exceed this radiation energy [61]. Thus the maximum brightness temperature that is possible is $T_{\rm B,max} = (mc^2/k)V_{\rm coh}n\zeta(\gamma - 1)$. Noting that $mc^2/k = 0.6 \times 10^{10}$ K, we can see that non-relativistic electrons cannot produce incoherent emission with $T_{\rm B} \gg 10^{10}$ K. Relativistic electrons can reach much higher temperatures but, for incoherent synchrotron emission, there is a limit at $\approx 10^{12}$ K, due to inverse Compton cooling [72], which sets the incoherent–coherent boundary. At higher temperatures the emission must be due to collective plasma processes with a large 'coherence factor' $nV_{\rm coh}$, with either negative absorption (masers) or emission by bunches. We now quickly review some incoherent and coherent emission mechanisms.

Cyclotron, gyrosynchrotron, synchrotron and curvature emission. A particle of charge q and mass m, travelling with a velocity $\mathbf{v}$ in a magnetic field $\mathbf{B}$, will experience a force $d/dt(\gamma m\mathbf{v}) = q\mathbf{v} \times \mathbf{B}$. Circular motion is described by $\mathbf{a} = \mathbf{\Omega} \times \mathbf{v}$ so we see that the particle will gyrate with a frequency $\omega_{\rm B}/\gamma$ where

$$\omega_{\rm B} = \frac{qB}{m}, \tag{1.2}$$

is the 'Larmor frequency'. For an electron, $\nu_{\rm B} = \omega_{\rm B}/2\pi = 2.8\ (B/\text{gauss})$ MHz. Solving Maxwell's equations for this scenario we find, as expected for an accelerating particle, that there is a radiation field.[2] When $\gamma \approx 1$ (i.e. non-relativistic) the resultant radiation is known as cyclotron radiation. It has a symmetric double-lobed beam shape in the direction of $+\mathbf{v}$ and $-\mathbf{v}$ and the radiation is at the frequency $\omega_{\rm B}$, i.e. the spectrum is a δ-function. The radiation pattern from a relativistic particle is beamed into a cone with half-angle $1/\gamma$ in the $+\mathbf{v}$ direction. Thus, the electric field observed is no longer sinusoidally varying, but instead consists of short pulses. As a result, for increasingly relativistic velocities,[3] the spectrum consists of more and more harmonics of $\omega_{\rm B}$. In the limit $v \rightarrow c$ the harmonics are contained within an envelope function as shown in Fig. 1.1. It is common to say that the radiation is concentrated at a 'critical frequency', $\omega_{\rm c} = \frac{3}{2}\gamma^3\omega_{\rm B}$. If an electron is subject to a large acceleration it will have a large $v_{\parallel}$ and thus a small pitch angle, so essentially moves along the field line. If the field line is curved, then the electron accelerates and therefore radiates what is known as curvature radiation. The associated critical

[1] The coherence volume is given by: $V_{\rm coh} = \lambda^3/\Delta\Omega(\Delta\omega/\omega)$, where λ is wavelength, $\Delta\omega/\omega$ is fractional bandwidth and $\Delta\Omega$ is the solid angle for the radiation. Note that $V_{\rm coh}n = 1$ for incoherent emission. Multiple particles in a coherence volume signals coherent emission.

[2] These are the Liénard–Wiechert fields, see e.g. Rybicki and Lightman [73], Chap. 3.

[3] The mildly relativistic regime is sometimes referred to as gyrosynchrotron emission.

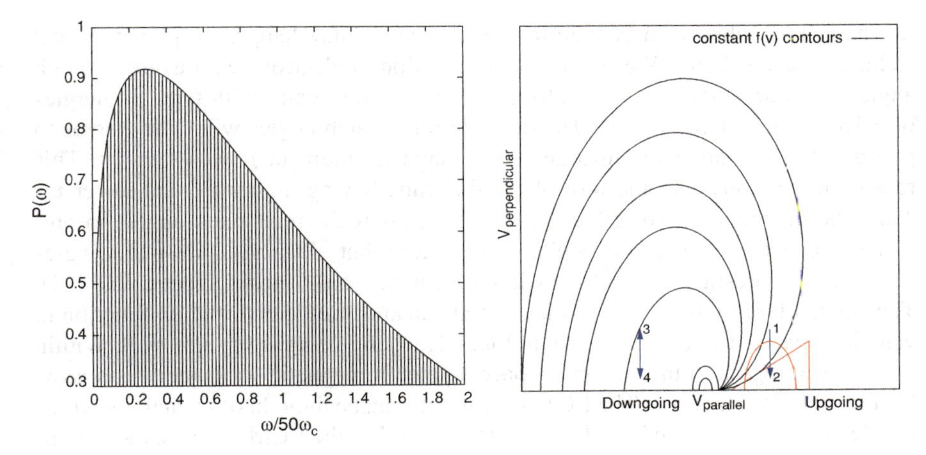

Fig. 1.1 (*Left*) The synchrotron spectrum of a single particle consists of many harmonics of ω_{B} within an envelope of the form $x\int_x^\infty K_{5/3}(x')dx'$, where $K_{5/3}$ is the modified Bessel function of the second kind, of order 5/3, and $x = \omega/\omega_{\mathrm{c}}$. In this example we have chosen $\omega_{\mathrm{c}} = 50\omega_{\mathrm{B}}$, corresponding to $v \sim 0.95c$. The total spectrum is a superposition of many such curves for particles with different velocities. (*Right*) Contours of the velocity–space distribution function. Conservation of the magnetic moment and/or magnetic mirroring creates the anisotropic distribution, in this example that of a 'loss cone' (*red triangle*). Electrons may transition from 1 → 2, which has a lower energy, but are unlikely to be replenished with electrons from 2 → 1 as there is a low density of particles there (i.e. a population inversion). This is not a problem on the down-going side (*points* 3 *and* 4) which is in equilibrium. The energy of the electron transitioning from 1 → 2 is lost to an electromagnetic wave, which, within a 'resonance ellipse' (*orange*), can grow to produce ECM emission

frequency can be written as $\omega_{\mathrm{c}} = \frac{3}{2}\gamma^3(v/2\pi\varrho)$, where ϱ is the radius of curvature of the field line.

Plasma Emission. Coherent plasma emission occurs indirectly. Instabilities in plasmas can generate turbulence in the form of longitudinal electron waves ('Langmuir waves'). Through secondary processes, e.g. scattering by different types of plasma waves, which are typically non-linear, the energy in these waves can be converted to radiation [62]. We do not discuss the details, as it is beyond our scope, except to say that the emission occurs at the plasma frequency

$$\nu_{\mathrm{p}} = \sqrt{\frac{ne^2}{\epsilon_0 m}}, \qquad (1.3)$$

the natural frequency of oscillation of a plasma, which for an electron is $\nu_{\mathrm{p}} = 8980\sqrt{n/\mathrm{cm}^{-3}}$ Hz, as well as its second harmonic, and the brightness temperatures can be $\lesssim 10^{17}$ K.

The electron cyclotron maser (ECM). The ECM can produce coherent non-thermal radiation if: (1) $\nu_{\mathrm{B}} \gg \nu_{\mathrm{p}}$, which requires a strong field and/or a low density plasma, and (2) a population inversion exists. These conditions can be met in the case of magnetic flux tubes with converging legs, such as may happen in planetary and stellar magnetospheres. We consider a distribution of electrons with isotropic pitch angles, $\theta = \arctan(v_\perp/v_\parallel)$, travelling in the magnetic field, with typical energies of ~10 keV to ~1 MeV [17]. The distribution of pitch angles will evolve so as to preserve the first adiabatic invariant—the magnetic moment $\mu = \frac{1}{2}mv_\perp^2/B$. This results in electrons near the foot of the flux tube having higher $v_\perp$ and lower $v_\parallel$. Also, the magnetic mirror effect causes electrons with small pitch angles to precipitate onto the planetary or stellar atmosphere, but electrons with pitch angles $\theta > \theta_{\mathrm{mirror}} = \arcsin(\sqrt{B_{\mathrm{top}}/B_{\mathrm{foot}}})$ will reflect in the converging magnetic field [21]. The net result of these two effects is to create an anisotropic distribution function in velocity–space, such as that shown in Fig. 1.1, where we have $\partial f/\partial p_\perp > 0$. Within an elliptical region in this velocity–space an electromagnetic wave mode can grow leading to ECM emission. The ECM growth rate can be thought of in analogy with a two-level laser where we have $\Gamma = d/dt(n_2/n_1)$. For the ECM we have the continuous distribution $f(\mathbf{p})$ so that this is more like $\Gamma \sim d/dt \int_{\mathbf{p}}(\partial f/\partial \mathbf{p})d^3\mathbf{p}$, which is positive if $\partial f/\partial p_\perp > 0$, i.e. we get a positive growth rate—a maser. If the electrons are accelerated due to the parallel component of an electric field, the low-velocity region of $v_\parallel - v_\perp$ space is forbidden [67], and the distribution is 'horseshoe' like, as is observed from in-situ measurements in the Earth's magnetosphere. The main differences in the resultant ECM emission from the two distributions are that the loss-cone emission is at slightly above ω_{B} and is stable for steady emission with $T_{\mathrm{B}} \lesssim 10^{14}$ K (as well as higher T_{B} bursts), whereas the horseshoe distribution is at slightly below ω_{B} and is stable for steady emission with $T_{\mathrm{B}} \lesssim 10^{20}$ K [18, 68]. In both cases the emission bandwidths are nominally quite narrow with $\Delta\omega/\omega \lesssim 0.01$, but, as this can happen at a variety of heights (and therefore B values and emission frequencies via Eq. 1.2), the ECM emission can persist to be broadband.

Pulsar Emission. The details of the pulsar emission mechanism are unknown, even 43 years after their initial discovery. As we will discuss below, the brightness temperatures of pulsar emission can exceed 10^{30} K, so the mechanism must be coherent, either as relativistic plasma radiation or a maser. Although many plasma instabilities and maser pumping mechanisms have been discussed in the literature [5, 22, 30, 56, 57], there is, as yet, no satisfactory consensus.

1.2 Transient Phase Space

Here, we define the 'transient phase space', after Cordes et al. [13]. This is simply the parameter space defined by L and νW, where $L = SD^2$ is the so-called (pseudo-)luminosity (in W Hz^{-1} or alternatively Jy kpc^2), S is the flux density, D is the distance to the source, ν is the frequency of the radiation and W is the width (in time) of the emission. This parameter space is plotted in Fig. 1.2. We can see from Eq. 1.1 that we can draw lines of *minimum* brightness temperature on this plane, with 10^{12} K marking the incoherent–coherent boundary. We can see that the known sources span a huge range both in νW and in L, but the majority of the plot

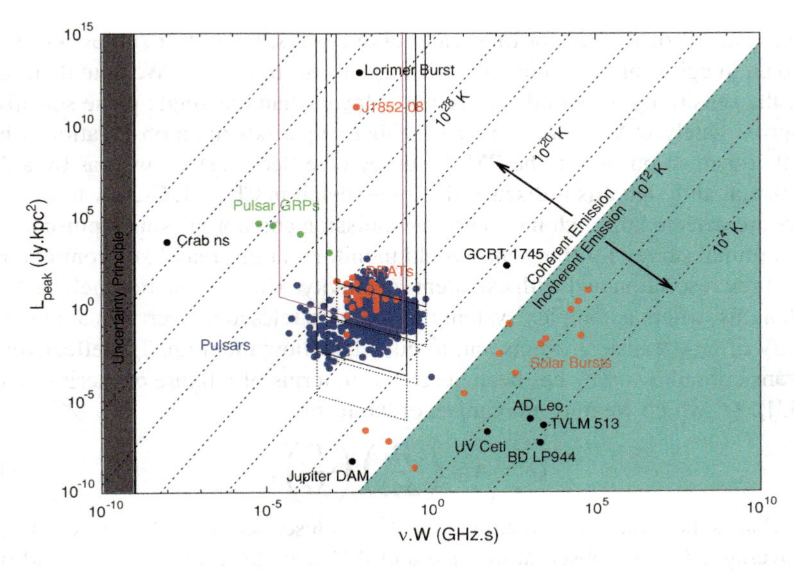

Fig. 1.2 The transient 'phase space' with known sources identified. The data in this plot are compiled from a number of sources, which, in addition to those mentioned in Sect. 1.3, include the ATNF Pulsar Catalogue [9, 34, 72]. We reiterate that lines of minimum brightness temperature are shown, e.g. the actual brightness temperature of the brown dwarf TVLM 513 is 2.4×10^{11} K and Jupiter DAM reaches 10^{20} K. The sensitivity of the PMSingle analysis (*black lines*), described in Chap. 4, to individual bursts, is overplotted for distances of 0.1, 1 and 10 kpc, respectively. With the effective area of the SKA the curves become lower by $\gtrsim$2 orders of magnitude in L (*dotted lines*). The LOFAR survey sensitivity curve (*pink line*) for a distance of 2 kpc is also shown

is empty. This sparse coverage illustrates our incomplete knowledge of the dynamic radio sky as well as the vast discovery space just waiting to be explored. It is the empty regions of this diagram that instruments like LOFAR, FAST and the SKA will be able to probe.

Overplotted on Fig. 1.2 are the sensitivity limits of the PMSingle analysis, as will be described in Chap. 4, to individual bursts, for source distances of 0.1, 1 and 10 kpc, respectively. Also plotted are the corresponding limits using the design specifications for the SKA (dotted lines), with $A_{\mathrm{eff}}/T_{\mathrm{sys}} = 2000\ \mathrm{m^2\,K^{-1}}$ [76] which are lower by approximately two orders of magnitude. In reality, the SKA curves will also be stretched to narrower pulse widths (to the left) and to wider pulses (to the right), in imaging surveys, as well to wider bandwidths (further to the right), by amounts limited only by computing power. The LOFAR sensitivity curves for likely pulsar survey parameters (B.W. Stappers, T. Hassall, private communication) is also shown for a distance of 2 kpc, out to which LOFAR is expected to give a complete census of Galactic neutron stars [82].[4] In fact the LOFAR curves will undoubtedly stretch

[4] Beyond this distance of ~2 kpc, the effects of interstellar scattering and dispersion will reduce LOFAR's sensitivity.

to higher values of νW, as for timescales above $\sim$1 s, the pulsar survey will make way to an imaging survey which will identify slower transients. We note that, in all cases, the sensitivity to periodic emission is lower than the single pulse sensitivity by approximately $\sqrt{N}$ where N is the number of periods in an observation, which is $\sim 10^2$ for most pulsars in the PMPS survey (see Sect. 4.3), as well as by a duty cycle factor, although this is discussed in more detail in Chap. 4. Clearly most of the pulsars are detected through their periodic emission and not via single pulses.

An optimal survey for transients would monitor a large area of sky continuously, at high time resolution and with excellent sensitivity. This is done routinely at X-ray wavelengths, where a complex system of alerts and telescope overrides are in place to notify of new transient events and to quickly follow them up. The effectiveness of a transient radio survey has been described in terms of a figure of merit (see e.g. [13, 33]), Q, which we must maximise, of the form:

$$Q = A_{\text{eff}} \left(\frac{\Omega}{\Delta\Omega}\right)\left(\frac{T}{\Delta t}\right), \tag{1.4}$$

where A_{eff} is the effective collecting area of the telescope/array, Ω is the solid angle sky coverage, T is the observation time and $\Delta\Omega$ and Δt are the angular and time resolutions, respectively. We can immediately see that Fig. 1.2 does not demonstrate a vital aspect of the LOFAR and SKA observations, their large value of $\Omega/\Delta\Omega$. While single-dish radio telescopes cannot have high sensitivity ($\propto A_{\text{eff}}$) combined with wide fields of view ($\propto 1/\sqrt{A_{\text{eff}}}$), LOFAR and SKA are arrays consisting of a large number of elements (dipoles in the case of LOFAR, dishes and aperture arrays in the case of the SKA) which can see a large fraction of the sky (large Ω) as well as being able to form numerous beams in software (small $\Delta\Omega$). The ability to continuously monitor the sky will enable the detection of the highest luminosity events, which may be quite rare (e.g. the 'Lorimer burst', described below, is the only known source at the highest luminosities). This is a key motivation in the transition from single dishes to large array 'software telescopes'.

1.3 Sources of Transient Radio Emission

The work undertaken for this thesis focuses on two main transient radio sources—pulsars and RRATs, both of which are discussed below and in the following chapters. However there are many other known radio transient sources, with several to be found in our own solar system, which are now described.

(i) *The sun*. The Sun is the brightest radio source in the sky with a flux of between $\sim$50–300 sfu varying on the 11-year solar cycle.[5] It exhibits a plethora of transient burst behaviours usually produced by large solar flares. Radio bursts ($\lesssim$300 MHz) are classed as Types I–V, with timescales ranging from a few seconds up to several

[5] 1 solar flux unit = 1 sfu = 10^{-22} W m^{-2} Hz^{-1} = 10^4 Jy. It is usually measured and quoted at $\lambda = 10.7$ cm, $\nu = 2.8$ GHz.

hours and weeks, all thought to be due to plasma emission [17]. In addition, there is microwave emission due to gyrosynchrotron (1–30 GHz) and ~10-ms ECM 'spike bursts' (~0.5–5 GHz). The associated brightness temperatures for these phenomena range from 10^7 to 10^{13} K.

(ii) *Planets*. All of the magnetised planets in our solar system—Earth, Jupiter, Saturn, Uranus and Neptune, show auroral radio emission (ARE) [84]. This emission arises from ECM instabilities in the planetary magnetospheres and the radiated power is found to be proportional to the incident solar wind power (this is known as 'Radio Bode's Law', Djorgovski and King [15]). The strongest ARE by far is that from Jupiter which has a magnetic field strength, $B = 4.3$ G, an order of magnitude stronger than the other planets. Jovian radio emission is broadband with three main components—the kilometre (KOM), hectometre (HOM) and decametre (DAM) components [84]. The DAM emission (at up to 40 MHz) itself consists of two components: Io-dependent and Io-independent. It is in the Io-dependent emission (which is independent of the solar wind) where we see strong radio bursts—the so-called 'Jovian S-bursts'[6] which can have $T_B \sim 10^{20}$ K. These bursts last for a few tens of milliseconds and originate from the polar regions on Jupiter at the base of the Io flux tube. This flux tube encloses a current of particles from Io to Jupiter due to the potential across Io's surface as it orbits in the magnetic field of Jupiter. Io essentially acts as a homopolar generator, aka unipolar inductor [23], and particles spewed from the highly volcanic moon end up flowing in this current and emit radio bursts due to the ECM instability. These bursts are narrow band (a few kHz), highly beamed and 100% polarised (RCP from northern latitudes on Jupiter, LCP from southern latitudes) and are strongly modulated by the rotation period of Jupiter (~10 hr).

Observing ARE has been proposed as a means of direct detection of exo-planets [51, 85]. The most recent blind search of several 100 nearby stars ($\lesssim$30 pc) has not detected anything, producing limits which are apparently 1–2 orders of magnitude higher than expected signals from 'Hot Jupiters' orbiting solar-like planets [52]). Of course a non-magnetic exoplanet could be detected if it had an Io-Jupiter-like interaction with its parent star. This requires a strong stellar magnetic field and so is usually discussed for magnetic white dwarf systems [83].

(iii) *Flare stars*. Radio emission is seen across the H–R diagram but the emission from the later spectral types (from F onwards) show predominantly non-thermal emission [63]. These cool stars with magnetic activity ($B \sim$ few kG) are all potential flare stars. 'Classical flare stars' are nearby red dwarfs (type M) that emit solar-like bursts both incoherent (~10 mJy on timescales of minutes to days) and highly-polarised coherent (~100 mJy on timescales of milliseconds to hours) flares. These flares are broadband and show highly variable structures in both frequency and time. Recently observations using the highest temporal and frequency resolutions have revealed 500 mJy bursts at L-band, of 2 ms duration from AD Leo [64]. In addition to the classical flare stars there are also close binaries containing main sequence stars—namely the RS CVn binaries, Algols and W Uma systems which show ~1 Jy flares with durations of 10–40 days [6, 71].

[6] This name is due to their S shape intensity curves in the frequency-time domain.

(iv) *Brown dwarfs*. Periodic radio emission from ultra-cool brown dwarfs has recently been reported [26, 28, 29]. This rotationally-modulated emission provides insight into the magnetic activity of dwarf stars. Stars with spectral type $\gtrsim$M3 are fully convective and thus conventional dynamos (like the $\alpha\Omega$ dynamo in the Sun) cannot operate. Despite this, magnetic mapping of such dwarfs has shown strong fields do exist [16]. The coherent radio bursts from brown dwarfs are due to coherent maser emission and confirm the existence of kG magnetic fields down to late M and L type dwarf stars [27]. As these bursts are rotationally modulated, from a star with a dipole field, they are in some sense 'slow pulsars'. It has been suggested that studies on these more manageable time-scales (their periods are $\sim$1–3 hr) might help in determining the unknown emission mechanism(s) in pulsars.

(v) *The 'Lorimer Burst'*. In 2007, the detection of a 30-Jy, 5-ms duration, highly dispersed ($DM = 375$ cm^{-3} pc) burst was reported [54]. The line of sight to the source has Galactic latitude $b = -41.8°$, which is so far out of the plane that the Galaxy should contribute a mere $DM_{\rm Gal} \approx 25$ cm^{-3} pc to the dispersion [14]. The lack of a counterpart in magnitude-limited galaxy catalogues [65] sets a lower-limit distance of $\sim$600 Mpc. Assuming an intergalactic medium (IGM) of fully ionised baryons gives a $DM_{\rm IGM}(z)$ relation [39, 40] which provides an upper-limit distance of $\sim$1 Gpc ($z \approx 0.3$). However considering the *DM* contribution from a putative host galaxy (if, say, it were similar to the Milky Way), the distance is taken to be $D \sim 500$ Mpc ($z \sim 0.12$) with considerable caveats due to the uncertainties in the free electron distributions in our own Galaxy, the IGM and any unknown source region, whose contribution is obviously unconstrained.

With 50 h of followup observations not showing any repeat of this phenomena it seems that it was a one-time event. Lorimer et al. [53] make a crude estimate of the 'event rate' for such bursts based on the duration of pulsar surveys that have been performed which have not detected such events and assuming an isotropic distribution on the sky. The rate so determined is $\sim$50 day^{-1} Gpc^{-1}, an order of magnitude higher than the estimated rates of Gamma Ray Bursts and binary neutron-star inspiral events but $\sim$20 times smaller than the core-collapse supernova rate in that volume. There are no recorded gamma-ray events or host galaxies at the burst's location and, having occurred in 2001, it was a pre-LIGO event.

(vi) *Ultra-high energy particles*. Ultra-high energy (UHE) cosmic rays with energies in the range 10^{18}–10^{20} eV are constantly incident upon the Earth's atmosphere. These impacts can result in nano-second radio bursts as intense as 10^6 Jy [35, 50]. However, identification of their source(s) is extremely difficult as cosmic magnetic fields deflect their paths so that the apparent distribution of these particles is isotropic on the sky. The highest energy particles are less deflected and some recent work has suggested that particles above 5.6×10^{19} eV may originate in active galactic nuclei [1]. Unfortunately, the flux of particles at such high energies is low and consequently the significance of the AGN correlation is questionable [3, 24]. An alternative method for determining UHE particle source regions is to observe astrophysical neutrinos produced from UHE particle interactions with CMB photons above $\sim 10^{20}$ eV [2]. This is appealing as neutrinos are not deflected by magnetic fields and cosmic rays above this so-called 'Greisen–Zatsepin–Kuzmin cutoff' have been observed [25].

Ongoing experiments aim to detect UHE neutrinos via the lunar Cherenkov method, i.e. observing the coherent Cherenkov emission (in the radio) from UHE particle interactions in the lunar regolith, although no detections have been made yet [41, 77].

(vii) *Pulsars and RRATs*. Neutron stars are the most populous member of the transient radio phase space. They are thus the most well studied. Neutron stars are highly compact stars with radii of ~10 km and masses of $\sim 1.4\mathrm{M}_{\odot}$ (see e.g. [78]). The gravitational potential energy felt by a test particle at the neutron star surface is 20% of its rest mass. There is thus extremely strong gravity to be considered. In addition we see that the average density is $\bar{\rho} = \mathrm{M}/[(4/3)\pi R^3] = 6.7\times10^{17}\ \mathrm{kg\,m^{-3}}$, more dense than nuclear matter. So, an advanced understanding of the strong force is needed in trying to determine the core composition. Outside the pulsar we have very strong magnetic fields with 10^{12} G being typical. These dominate outside the star and an understanding of the electrodynamics of non-neutral plasmas seems to be needed to explain the physics of neutron star magnetospheres [60]. Including the super-fluids and super-conductors in the interior we can see that neutron stars are ideal laboratories for many and most areas of extreme physics.

Qualitatively, pulsars are those rapidly rotating neutron stars which emit an apparently steady, narrow beam of emission. This emission seems to originate from fixed regions on/above the neutron star surface so that, modulated by the star's rotation, the beams can sweep across our line of sight and be detected at Earth. Thus observers see a 'pulse' of emission once per rotation [44]. Rotating radio transients (RRATs) are those rapidly rotating neutron stars whose emission does not seems to be steady and they are usually seen to be 'off'. For RRATs, we see a pulse when the emission beam cuts our line of sight if it happens to be 'on'. The brightest pulses detected from a neutron star are the giant pulses from the Crab pulsar, 10^4 Jy at 1.4 GHz [43], and 10^5 Jy at 400 MHz [31]. In addition to their pulsed emission, neutron stars can be transient in other respects—there are pulsars known in eclipsing binaries and there are 'intermittent pulsars', which turn on and off for weeks at a time [49], and (see Sect. 8.2). Neutron stars, pulsars and RRATs are discussed in detail in Chap. 2. The study of RRATs in particular is the main focus of the work presented in this thesis.

(viii) *Miscellaneous sources*. GCRT J1745−3009 is a transient radio source detected in archival VLA monitoring observations of the Galactic Centre at 0.33 GHz [37]. Five 1-Jy, 10-min bursts were observed consistent with a periodicity of 77.012 ± 0.021 min [80]. The emission mechanism seems to be coherent but due to distance uncertainties the brightness temperature lies somewhere in the range $10^{12} - 10^{16}$ K. Numerous explanations have been put forward, which include flaring magnetic brown dwarfs [73] and white dwarf pulsars [86]. Another, apparently similar source, GCRT J1742−3001 has recently been reported [38].

Other sources of transient emission include the radio bursts from X-ray binaries,[7] e.g. the 20-Jy bursts seen in Cygnus X-3 [19, 20]. Radio variability and bursting is observed in Active Galactic Nuclei at millimeter and centimeter wavelengths [4, 61, 62]. OH Maser emission has been observed to flare, e.g. in Cepheus A

[7] Radio emitting X-ray binaries are sometimes referred to as 'microquasars'.

an increase in flux density by a factor of 250, to 25 Jy, has been observed over timescales of a few months [10]. In the past 4 years, 3 radio-emitting magnetars have been discovered [7, 8, 53]. These sources, which had been thought to be radio-quiet, have exhibited transient pulsar-like behaviour. Another important source of transient radio behaviour is intra-day variability (IDV). IDV is the flickering in flux density seen in numerous flat-spectrum extragalactic radio sources [44]. IDV is thought to be caused by interstellar scintillation with typical amplitudes of a few percent although much larger IDV changes have been observed [42]. IDV will be a source of worry when deciding on flux calibrators for the SKA [13].

(ix) *Theoretical sources*. In addition to the known sources there are many proposed sources of astrophysical radio bursts. For instance a system of in-spiralling NSs is predicted to have a coherent radio pre-cursor burst a few seconds before merger and the expected gravitational wave burst. This signal is predicted to have a radio flux of $\sim$2 mJy$(100\ \mathrm{Mpc}/\mathrm{D})^2$ at 400 MHz [36]. Radio bursts are also predicted from annihilating black holes. Black holes emit exact blackbody radiation[8] at the Hawking temperature $T_{\mathrm{H}} \approx 20(\mathrm{M}_{\oplus}/M)$ mK [32]. As they lose energy they lose mass and are thus predicted to evaporate to some critical mass at which point the rest of their energy is emitted in a fire-ball of electrons and positrons. The energy of the fire-ball is predicted to be observable as an electromagnetic millisecond burst in the radio region of the spectrum [70]. Such a scenario is proposed for so called mini-holes (i.e. primordial black holes), as the evaporation time-scale is much larger than the age of the universe for stellar mass black holes. Expanding supernova shells of conducting material which interact with the star's pre-existing magnetic field can result in broadband coherent transient ($<$1 s) bursts in the radio [11]. Searches for these events have been unsuccessful [66], but such exotic scenarios have been proposed as explanations of the Lorimer Burst. It seems that a firm confirmation of such events require multi-wavelength detections and possibly an associated gravitational wave signal. SETI networks also search for transient radio signals [14, 47, 48] using dedicated Arecibo Telescope observations via its distributed computing project SETI@home[9] which has over 1 million users and 2.5 million computers contributing. Recent developments in the search for broadband pulses include the Astropulse survey at Arecibo and the Fly's Eye experiment which uses the Allen Telescope Array (ATA) in California [79]. Gamma Ray Bursts (GRBs) are also proposed to have associated prompt radio bursts of $\sim$100 s duration at frequencies of tens of MHz with highly uncertain flux density estimates with an upper-limit of $\sim$100 Jy [75, 81] but searches for such events have not yet been successful [45, 46].

---

[8] Unlike thermal radiation from blackbodies which only obey Planck's law statistically, i.e. on average. Whereas statistical blackbody radiation contains information on the emitting object, Hawking radiation does not, depending only on M_{BH} (and J_{BH} and Q_{BH}).

[9] http://setiathome.ssl.berkeley.edu

1.4 Thesis Outline

This thesis is organised into the following chapters:

- Chapter 2 discusses the physics of neutron stars and the pulsar model. Presented also is a review of our knowledge of RRATs as it stood in 2007, at the beginning of the research for this thesis.
- Chapter 3 investigates what the implications are, if RRATs, and indeed other known neutron star classes, are in fact isolated Galactic populations.
- Chapter 4 describes the methodology and presents the results of the PMSingle re-analysis of the Parkes Multi-beam Pulsar Survey which was performed with the aim of finding new RRAT sources.
- Chapter 5 outlines the methods for, and difficulties associated with, timing observations of RRATs and presents results from observations with the Lovell Telescope.
- Chapter 6 presents the timing solutions for the newly discovered sources from the PMSingle analysis.
- Chapter 7 describes a planned X-ray observation as well as searches for optical bursts in the RRAT J1819−1458.
- Chapter 8 is an overview of the work of other authors in the past 3 years.
- Chapter 9 discusses conclusions we have arrived upon as a result of this research.
- Appendix A useful radio astronomy equations.
- Appendix B supplementary information for Chap. 2.
- Appendix C supplementary information for Chap. 3.
- Appendix D supplementary information for Chap. 4.
- Appendix E supplementary information for Chap. 5.

References

1. J. Abraham et al., Science **318**, 938 (2007)
2. J. Abraham et al., Phys. Rev. Lett. **101**, 061101 (2008)
3. P. Abreu et al., ArXiv e-prints (2010) (astro-ph/1009.1855)
4. H.D. Aller, M.F. Aller, G.E. Latimer, P.E. Hodge, ApJS **59**, 513 (1985)
5. E. Asseo, G. Pelletier, H. Sol, MNRAS **247**, 529 (1990)
6. T.S. Bastian, Space Science Reviews, **68**, 261 (1994)
7. F. Camilo, S.M. Ransom, J.P. Halpern, J. Reynolds, D.J. Helfand, N. Zimmerman, J. Sarkissian, Nature **442**, 892 (2006)
8. F. Camilo, J. Reynolds, S. Johnston, J.P. Halpern, S.M. Ransom, ApJ 681 (astro-ph/0802.0494) (2008)
9. I. Cognard, J.A. Shrauner, J.H. Taylor, S.E. Thorsett, ApJ **457**, L81 (1996)
10. R.J. Cohen, G.C. Brebner, MNRAS **216**, 51 (1985)
11. S.A. Colgate, P.D. Noerdlinger, ApJ **165**, 509 (1971)
12. J.M. Cordes, T.J.W. Lazio, preprint (arXiv:astro-ph/0207156)
13. J.M. Cordes, T.J.W. Lazio, M.A. McLaughlin, New Astronomy Review **48**, 1459 (2004)
14. J.M. Cordes, T.J.W. Lazio, C. Sagan, ApJ **487**, 782 (1997)
15. S. Djorgovski, I.R. King, ApJ **277**, L49 (1984)

16. J. Donati, T. Forveille, A.C. Cameron, J.R. Barnes, X. Delfosse, M.M. Jardine, J.A. Valenti, Science **311**, 633 (2006)
17. G.A. Dulk, Ann. Rev. Astr. Ap **23**, 169 (1985)
18. R.E. Ergun, C.W. Carlson, J.P. McFadden, G.T. Delory, R.J. Strangeway, P.L. Pritchett, ApJ **538**, 456 (2000)
19. R.P. Fender, S.J. Bell Burnell, E.B. Waltman, Vistas Astron. **41**, 3 (1997)
20. R.P. Fender, S.J. Bell Burnell, E.B. Waltman, G.G. Pooley, F.D. Ghigo, R.S. Foster, MNRAS **288**, 849 (1997)
21. R. Fitzpatrick, Introduction to Plasma Physics: a graduate level course (2008)
22. V.L. Ginzburg, V.V. Zheleznyakov, Comm. Astrophys. **2**, 197 (1970)
23. P. Goldreich, D. Lynden-Bell, ApJ **156**, 59 (1969)
24. D. Gorbunov, P. Tinyakov, I. Tkachev, S. Troitsky, Sov. J. Exp. Theor. Phys. Lett. **87**, 461 (2008)
25. K. Greisen, Phys. Rev. Lett. **16**, 748 (1966)
26. G. Hallinan, A. Antonova, G.J. Doyle, S. Bourke, W.F. Brisken, A. Golden, ApJ **653**, 690 (2006)
27. G. Hallinan, A. Antonova, J.G. Doyle, S. Bourke, C. Lane, A. Golden, ApJ **684**, 644 (2008)
28. G. Hallinan et al., Mem. S. A. It **78**, 304 (2007a)
29. G. Hallinan et al., ApJ **663**, L25 (2007b)
30. T.H. Hankins, J.S. Kern, J.C. Weatherall, J.A. Eilek, Nature **422**, 141 (2003)
31. T.H. Hankins, B.J. Rickett, in Methods in Computational Physics—Radio Astronomy, vol. 14 (Academic Press, New York, 1975), p. 55
32. S.W. Hawking, Nature **248**, 30 (1974)
33. J.W.T. Hessels, B.W. Stappers, A.G.J. van Leeuwen, ArXiv e-prints (2009) (astro-ph/0903.1447)
34. G. Hobbs, R. Manchester, A. Teoh, M. Hobbs, in *Young Neutron Stars and Their Environments*, vol. 1, ed. by F. Camilo, B.M. Gaensler, IAU Symposium 218 (Astronomical Society of the Pacific, San Francisco, 2004), p. 139
35. T. Huege, H. Falcke, A&A **412**, 19 (2003)
36. D. Hutsemékers, H. Lamy, A&A **367**, 381 (2001)
37. S.D. Hyman, T.J.W. Lazio, N.E. Kassim, P.S. Ray, C.B. Markwardt, F. Yusef-Zadeh, Nature **434**, 50 (2005)
38. S.D. Hyman, R. Wijnands, T.J.W. Lazio, S. Pal, R. Starling, N.E. Kassim, P.S. Ray, ApJ **696**, 280 (2009)
39. S. Inoue, MNRAS **348**, 999 (2004)
40. K. Ioka, ApJ **598**, L79 (2003)
41. C.W. James, R.D. Ekers, J. Álvarez-Muñiz, J.D. Bray, R.A. McFadden, C.J. Phillips, R.J. Protheroe, P. Roberts, Phys. Rev. D **81**, 042003 (2010)
42. D.L. Jauncey et al., Ap&SS **278**, 87 (2001)
43. R. Karuppusamy, B.W. Stappers, W. van Straten, **515**, A36 (2010)
44. L.L. Kedziora-Chudczer, D.L. Jauncey, M.H. Wieringa, A.K. Tzioumis, J.E. Reynolds, MNRAS **325**, 1411 (2001)
45. D.M. Koranyi, D.A. Green, P.J. Warner, E.M. Waldram, D.M. Palmer, MNRAS **271**, 51 (1994)
46. D.M. Koranyi, D.A. Green, P.J. Warner, E.M. Waldram, D.M. Palmer, MNRAS **276**, L13 (1995)
47. E.J. Korpela et al., in *Astronomical Society of the Pacific Conference Series*, Vol. 420, ed. by K.J. Meech, J.V. Keane, M.J. Mumma, J.L. Siefert, D.J. Werthimer, SETI with Help from Five Million Volunteers: The Berkeley (2009), p. 431
48. M. Kramer, A.G. Lyne, J.T. O'Brien, C.A. Jordan, D.R. Lorimer, Science **312**, 549 (2006)
49. G.I. Langston, R. Bradley, T. Hankins, B. Mutel, Nucl. Instrum. Methods Phys. Res. A **604**, 116 (2009)
50. J. Lazio et al., in *Astronomy*, Vol. 2010, AGB Stars and Related Phenomena, Astro2010: The Astronomy and Astrophysics Decaded Survey (2009), p. 177

51. T.J.W. Lazio, S. Carmichael, J. Clark, E. Elkins, P. Gudmundsen, Z. Mott, M. Szwajkowski, L.A. Hennig, AJ **139**, 96 (2010)
52. L. Levin et al., ArXiv e-prints (2010) (astro-ph/1007.1052)
53. D.R. Lorimer, M. Bailes, M.A. McLaughlin, D.J. Narkevic, F. Crawford, Science **318**, 777 (2007)
54. D.R. Lorimer, M. Kramer, *Handbook of Pulsar Astronomy* (Cambridge University Press, 2005)
55. M. Lyutikov, R.D. Blandford, G. Machabeli, MNRAS **305**, 338 (1999)
56. D. Melrose, in *Young Neutron Stars and Their Environments*, Vol. 1, IAU Symposium 218. (Astronomical Society of the Pacific, San Francisco, 2004), p. 349
57. D.B. Melrose, Ann. Rev. Astr. Ap **29**, 31 (1991)
58. D.B. Melrose, in *IAU Symposium*, Vol. 257, ed. by N. Gopalswamy, D.F. Webb, IAU Symposium (2009), p. 305
59. F.C. Michel, *Theory of Neutron Star Magnetospheres* (University of Chicago Press, Chicago, 1991)
60. C.G. Mundell, P. Ferruit, N. Nagar, A.S. Wilson, ApJ **703**, 802 (2009)
61. E. Nieppola, T. Hovatta, M. Tornikoski, E. Valtaoja, M.F. Aller, H.D. Aller, AJ **137**, 5022 (2009)
62. R.A. Osten, ArXiv e-prints (astro-ph/0801.2573) (2008)
63. R.A. Osten, T.S. Bastian, ApJ **674**, 1078 (2008)
64. G. Paturel, C. Petit, P. Prugniel, G. Theureau, J. Rousseau, M. Brouty, P. Dubois, L. Cambrésy, A&A **412**, 45 (2003)
65. S. Phinney, J.H. Taylor, Nature **277**, 117 (1979)
66. P.L. Pritchett, R.J. Strangeway, J. Geophys. Res. **90**, 9650 (1985)
67. P.L. Pritchett, R.J. Strangeway, C.W. Carlson, R.E. Ergun, J.P. McFadden, G.T. Delory, J. Geophys. Res. **104**, 10317 (1999)
68. A.C.S. Redhead, ApJ **426**, 51 (1994)
69. M.J. Rees, Nature **266**, 333 (1977)
70. M.T. Richards, E.B. Waltman, F.D. Ghigo, D.S.P. Richard, ApJS **147**, 337 (2003)
71. R. Romani, S. Johnston, ApJ **557**, L93 (2001)
72. S. Roy, S.D. Hyman, S. Pal, T.J.W. Lazio, P.S. Ray, N.E. Kassim, ArXiv e-prints (astro-ph/1001.5394) (2010)
73. G.B. Rybicki, A.P. Lightman, Radiat. Process. Astrophys. (Wiley, New York, 1979)
74. A. Sagiv, E. Waxman, ApJ **574**, 861 (2002)
75. R.T. Schilizzi, P. Alexander, J.M. Cordes, SKA draft (2007), http://www.skatelescope.org/PDF/PreliminarySpecificationsoftheSquareKilometreArray_v2.7.1.pdf
76. O. Scholten et al., Phys. Rev. Lett. **103**, 191301 (2009)
77. S.L. Shapiro, S.A. Teukolsky, *Black Holes, White Dwarfs and Neutron Stars*. The Physics of Compact Objects (Wiley, New York, 1983)
78. A. Siemion et al., ArXiv e-prints (astro-ph/0811.3046) (2008)
79. H. Spreeuw, B. Scheers, R. Braun, R.A.M.J. Wijers, J.C.A. Miller-Jones, B.W. Stappers, R.P. Fender, A&A **502**, 549 (2009)
80. J. Tarter, Ann. Rev. Astr. Ap **39**, 511 (2001)
81. V.V. Usov, J.I. Katz, A&A **364**, 655 (2000)
82. J. van Leeuwen, B.W. Stappers, A&A **509**, A7 (2010)
83. A. Wiles, K. Wu, A&A **432**, 1091 (2005)
84. P. Zarka, J. Geophys. Res. **20**, 159 (1998)
85. P. Zarka, R.A. Treumann, B.P. Ryabov, V.B. Ryabov, Ap&SS **277**, 293 (2001)
86. B. Zhang, J. Gil, ApJ **631**, L143 (2005)

Chapter 2
Neutron Stars

Oh my God ... it has finally happened, he has become so massive that he collapsed into himself like a neutron star

Stewie Griffin

$$\rho_{\mathrm{NS}} V_{\mathrm{teaspoon}} = 5.5 \times 10^{8} M_{\mathrm{T-Rex}}$$

It is now understood that sufficiently massive stars will end their lives violently with explosions which can outshine their host galaxy. The core, of the star that was, collapses into a super-dense ball with exotic properties. These objects are known as neutron stars, the most exciting physical laboratories that nature has provided us with. Here we describe the various avenues of stellar evolution before focusing on the neutron star end-point. We describe the structure of the re-born zombie stellar remnant and its manifestation as a radio source. The physics is extreme, the environments are deadly, zombies are cool.

2.1 The Life and Death of Stars

Before discussing stellar remnants, it makes sense to recap the life of main sequence (MS) stars. 'Proto-stars' are born in molecular clouds and before contracting along their respective Hayashi (and Henyey[1]) tracks. During this contraction the star heats and powers itself by releasing gravitational potential energy. This can be seen from the Virial Theorem which, in its simplest form,[2] is $2U + \Omega = 0$ where U is the star's internal energy and Ω its gravitational potential energy. Thus stars are commonly said

---

[1] A star with $M > 0.5\mathrm{M}_{\odot}$ will move at essentially constant luminosity along a horizontal line on the Hertzsprung-Russell diagram after it has travelled down its Hayashi track.

[2] More generally, for a rotating body, the Virial Theorem is $\frac{1}{2}\frac{d^2 I}{dt^2} = 2U + \Omega + 2T + M_{\mathrm{B}} +$ (surface terms), where I is the moment of inertia, T is kinetic energy in bulk motion and M_{B} is any magnetic energy.

E. F. Keane, *The Transient Radio Sky*, Springer Theses,
DOI: 10.1007/978-3-642-19627-0_2,

to have a negative heat capacity. This is a general feature of any system dominated by self-gravity.

Nuclear burning thresholds depend strongly upon temperature and density, e.g. $\epsilon_{pp} \propto \rho T^4$ for the p–p chain where ϵ_{pp} is the rate of energy produced per unit mass.[3] If and when it becomes hot and dense enough in the stellar core nuclear burning will occur. The point when this starts is known as the zero-age main sequence (ZAMS) line. At this point the future evolution of the star into a compact object is inevitable and is determined by its initial mass. Stars wherein nuclear burning does not begin are known as 'failed stars', or 'brown dwarfs'. They do not possess sufficient initial mass and their cores never get hot and dense enough for nuclear burning to occur. The brown dwarf/stellar boundary is hazy but lies at $\sim 80\ M_{♃} \approx 0.08\ M_{\odot}$. The brown dwarf/planet boundary is more well defined at $13\ M_{♃} \approx 0.01\ M_{\odot}$.

For a star with a given initial mass, we can determine what type of stellar remnant it will become once its MS lifetime has ended [60, 65]. To do this we must know the initial mass function (IMF). The IMF, denoted by $\Phi(m)$, is a probability density function such that $\Phi(m)dm$ is the probability that a star will be born with an initial mass in the range of $m + dm$. Current evidence suggests that there is an intrinsic Galactic IMF with variations consistent with the expected statistical variations [43] although the universality of the IMF is a hotly debated topic of research. The observed IMF can be modelled as a piece-wise combination of three power-laws with the high-mass slope ($\Phi \propto m^{-2.35}$) agreeing with the original IMF calculation of Salpeter [59]. The IMF of Kroupa [30] is currently the best determined ([43], S. Stahler, private communication) which we will refer to as IMF-3. However there is a subtlety we must note which means that the observed and intrinsic IMFs may not be the same due to the effect of un-resolved binaries [31][4] so that we need to consider an IMF consisting of a combination of four power-laws, IMF-4. We can write both of these scenarios like:

$$\Phi(m) = \begin{cases} K_1 \cdot (m)^{\alpha_1} & m_1 < m < m_2 \\ K_2 \cdot (m)^{\alpha_2} & m_2 < m < m_3 \\ K_3 \cdot (m)^{\alpha_3} & m_3 < m < m_4 \\ K_4 \cdot (m)^{\alpha_4} & m_4 < m < m_{\max} \end{cases} \tag{2.1}$$

where the K terms are constants and IMF-3 has $\alpha_3 = \alpha_4$. IMF-3 has $\alpha_1 = -0.3, \alpha_2 = -1.3, \alpha_3 = -2.3$ (Salpeter-like). IMF-4 is of the same form but with $\alpha_4 = -2.7$ (Scalo-like). The mass values are $m_1 = 0.01\ M_{\odot}, m_2 = 0.08\ M_{\odot}, m_3 = 0.5\ M_{\odot}, m_4 = 1\ M_{\odot}$ and we use a maximum mass of $120\ M_{\odot}$. The constants are determined from continuity and by requiring normalisation, i.e.

[3] Later nuclear reaction thresholds depend even more strongly on temperature: for the CNO cycle $\epsilon_{CNO} \propto \rho T^{16}$; for the triple-$\alpha$ process $\epsilon_{3\alpha} \propto \rho^2 T^{40}$.

[4] This can result in unresolved stars being mistaken for one, more massive, star so that the derived probability at higher masses will be artificially increased.

$$\int_{0.01}^{120} \Phi(m)dm = 1 \tag{2.2}$$

We note that the low-mass end of the IMF (for low mass red dwarfs and brown dwarfs) is quite uncertain due to the difficulty in measuring the luminosities of these faint stars.[5]

Stars with *final masses* below the Chandrasekhar limit, $M_{\rm Ch} = 1.43(2/\mu_{\rm e})^2\ {\rm M}_\odot$, will become white dwarfs, where $\mu_{\rm e}$ is the average number of nucleons per electron which equals 2 in the fully ionised case. There is a similar, but less well-determined, limit for neutron star progenitors, which is known as the Oppenheimer-Volkoff (OV) limit which lies somewhere in the range 1.4–3 $M_\odot$. Stars with final masses above the Chandrasekhar limit but below the OV limit will become neutron stars. Stars with final masses above the OV limit will become black holes. With known IMF and a known star-formation rate we can calculate the fraction of stars which will end up as white dwarfs, neutron stars and black holes, if we know the map between initial and final mass distributions. Herein lies the difficulty as this map depends on mass loss during MS evolution, something which is very difficult to model. In general this map also depends on the star's initial conditions (e.g. metallicity, binarity) and environment. As such the transition masses are uncertain. The initial mass range usually taken for stars to form neutron stars (NSs) via core-collapse SN^e is 8.5–25 $M_\odot$, but recently it has been argued [68] that the fraction of stars which will undergo Fe core-collapse SN^e are those born in the mass range of $11 \pm 1 - 25\ M_\odot$. Single stars in the range of $8 - 11 \pm 1\ M_\odot$ are expected to leave behind degenerate O–Ne–Mg cores as white dwarfs (not neutron stars) due to heavy wind mass loss during their AGB phase [52, 68]. Binary stars in this mass range may undergo another type of SN which can produce neutron stars, an electron capture SN, if they are in interacting (close-binary) systems [52] but this is a very small proportion of neutron stars. We discuss the WD-NS transition mass region in more detail in Sect. 3.5. We take 11 $M_\odot$ as the WD-NS transition mass and 25 $M_\odot$ as the NS-BH transition mass [18] and calculate the fraction of (non-failed) stars which will end up as WDs, NSs and BHs like:

$$f_{\rm WD} = \int_{0.08}^{11} \Phi(m)dm, \quad f_{\rm NS} = \int_{11}^{25} \Phi(m)dm, \quad f_{\rm BH} = \int_{25}^{120} \Phi(m)dm. \tag{2.3}$$

The estimates this provides are listed in Table 2.1.

We can see that 99% of stars will end their lives as WDs. Less than 1% of stars will end up as neutron stars and fewer still will become black holes. As we will discuss in Chap. 3, if we know the Galactic star-formation rate we can go further and estimate birthrates for these different populations. For now we discuss the properties of degenerate matter and the structure of neutron stars, although the initial part of our description applies to white dwarfs also.

---

[5] Recall the empirical relation for MS stars which states that $L \propto M^{\sim 3.5}$ [54].

Table 2.1 Stellar Remnants: outcome of stellar evolution of isolated, non-failed stars for given initial mass ranges

Remnant	Initial mass range ($M_\odot$)	Fraction (IMF-3) (%)	Fraction (IMF-4) (%)
White dwarf	0.08–11	99.6	99.9
Neutron star	11–25	0.3	0.1
Black hole	25–120	0.1	0.01

2.2 Theory of Neutron Stars

2.2.1 Neutron Stars on the Back of an Envelope

As we have mentioned above, stars with initial masses in the range ~11–25 $M_\odot$ will, once they have burned all of their nuclear fuel (so that they have an ^{56}Fe core), undergo a Type II supernovae whereby the outer layers of the star are expelled in a hugely energetic explosion. The highly compact core which remains is the (proto-)neutron star (NS). But how can we know anything about this compact NS remnant? Fortunately several manifestations of NSs (which we will describe below) are observable, spanning the entire electromagnetic spectrum from radio waves to γ-rays. By observing sources in binary systems, we can determine the masses of NSs, e.g. simple Keplerian dynamics gives us the binary mass function and a measurement of relativistic periastron advance gives us the total mass $M = M_{\rm PSR} + M_{\rm companion}$ as $\dot{\omega} \propto M^{2/3}$ [37]. Figure 2.1 shows all of the NS masses which have been determined and we can see that, although there is a span of values in the range 1–2.8 $M_\odot$, most values lie within a narrow mass range around 1.4 $M_\odot$ with a median mass of 1.38 $M_\odot$. Typically 1.4 $M_\odot$ is taken as the canonical NS mass.

Determining a canonical radius is more complicated as the full equation of state (i.e. at all densities) is required and this is unknown. Using simple arguments however we can quickly place limits on the NS radius. If we require that the NS not spin faster than its break-up speed, i.e. the surface velocity of a particle at the equator be less than the Keplerian orbital velocity, we get $\omega_{\rm max} = \sqrt{GM/R_{\rm max}^3}$ which yields a maximum radius of $R_{\rm max} = 16.8$ km $(M/1.4\ M_\odot)^{1/3}(P/{\rm ms})^{2/3}$. For the fastest spinning pulsar (PSR J1748−2446ad, $P = 1.39$ ms, [20]) this gives $R_{\rm max} = 20.9$ km. We can determine a lower limit for the radius by considering a star of constant density in General Relativity. By requiring the central pressure be finite we get $R_{\rm min} = (9/8)R_{\rm S}$, where $R_{\rm S} = 2GM/c^2$ is the Schwarzschild radius (see Appendix B). Considering a range of more realistic equations of state, Glendenning [15] found $R_{\rm min} \sim 1.5\ R_{\rm S} = 6.2\,{\rm km}\,(M/1.4\ M_\odot)$. Further constraints can be determined from X-ray observations, e.g. measuring the gravitational redshift of absorption lines to get an estimate for the 'compaction' factor M/R (see e.g. Cottam et al. [13]), or through observations of the thermal emission from neutron stars in globular clusters, where the distance is accurately known (see e.g. Bogdanov et al. [8]). In the next section we will discuss the neutron star interior, but for now we note that more detailed

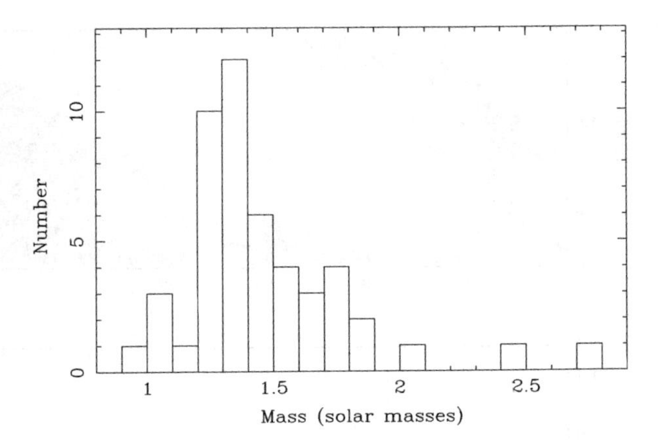

Fig. 2.1 A histogram of the measured masses of neutron stars, based on data published in Lorimer [35], an update of earlier work by Stairs [66]

considerations of the NS interior structure yield radii in the range 10–12 km, and 10 km is usually taken as the canonical NS radius.

2.2.2 Neutron Star Structure

We have strong reason to believe that NSs are spherical—if they were not they would emit detectable gravitational waves. The most recent LIGO limit gives upper limits on neutron star ellipticities of 10^{-6} [1]. To determine the structure of a NS we thus need to solve the hydrostatic equilibrium equations of a spherical star in General Relativity, the Tolmann-Oppenheimer-Volkoff' (TOV) Equations [60],

$$\frac{dm}{dr} = 4\pi r^2 \rho, \tag{2.4}$$

$$\frac{dP}{dr} = \frac{(\rho + P)(m + 4\pi r^3 P)}{r(r - 2m)}, \tag{2.5}$$

where we have set $G = c = 1$ and $M = \int_0^R 4\pi r^2 \rho(r) dr$ is to be interpreted as the mass of the star *plus* its negative binding energy—its 'gravitational mass', $\int_0^R 4\pi r^2 \rho(r) \sqrt{g_{rr}} dr$ gives the 'baryonic mass'. The TOV equations can be solved numerically but require, as input, an equation of state (EoS), $P = P(\rho)$, and herein lies the difficulty in determining NS internal structure. The solution for a non-relativistic gas of neutrons is shown in Fig. 2.2, a reproduction of the result obtained by Oppenheimer and Volkoff [48].

How do we determine the correct EoS for a NS? With a mass of 1.4 $M_\odot$, crammed into a radius of just under three times its Schwarzschild radius, we can see that we are dealing with an object with average density $\bar{\rho} = 6.7 \times 10^{17}$ kg m^{-3}, which is more dense than nuclear matter! Thus, we need the EoS of degenerate matter at densities up to and much higher than anything achievable in terrestrial laboratories.To do

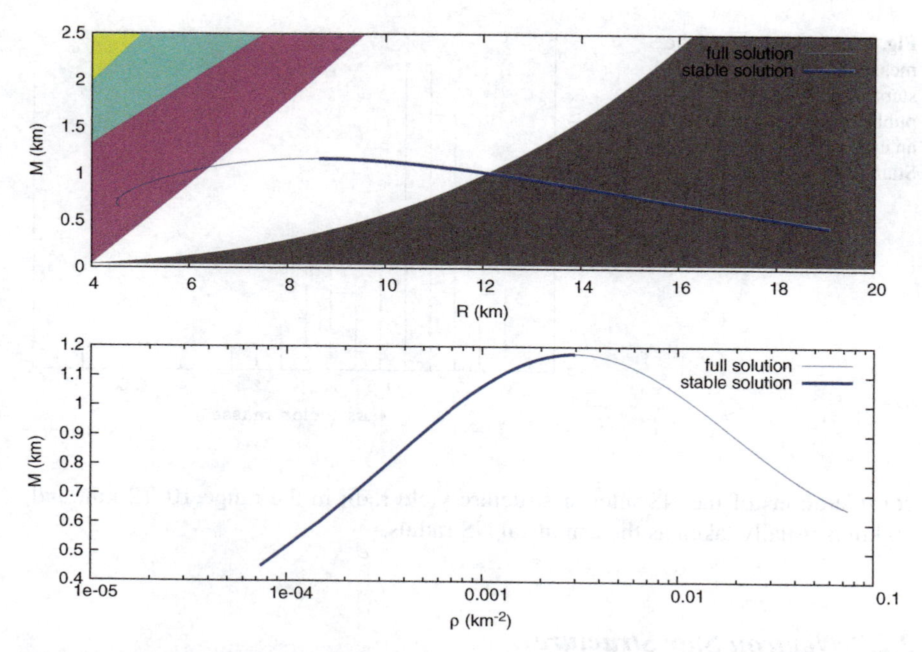

Fig. 2.2 (*Top*) The mass-radius curve for a simple EoS considering only non-relativistic degenerate neutrons. The thin *blue line* shows the solution over a wide range of radii and densities. The thick *blue line* denotes the stable solution, where $dM/d\rho_c > 0$. Areas excluded by GR (*yellow*), causality (*sky blue*), glitch measurements (*pink*) and rotation (*grey*) are plotted. Although not shown here, there are a plethora of other EoS which fill the allowed region of this mass-radius space (see e.g. Lattimer and Prakash [32, 33]). (*Bottom*) The mass-central density curve for the same solution. All quantities are in geometrised units ($G = c = 1$)

this we note that the density will have some radial profile decreasing outwards. We can determine the interior structure starting from the NS crust and working our way towards the centre. *We make the assumption* that, at a given density, the configuration is the one that minimises the energy density. We also take it that this equilibrium configuration arises in a short time, i.e. that the transition from 'proto-neutron star' to NS is quick. What happens in the interim period when the NS cools very quickly, down to temperatures of $\sim 10^6$ K, will not be considered here.

Degenerate Gases

At low densities the equilibrium nucleus is $^{56}_{26}$Fe, of which the pre-collapse stellar core was entirely comprised of. Post-collapse the remnant interior is much denser so that we have degenerate matter. At the crust we have iron nuclei and freedegenerate electrons. Deeper down (as we will show) we will get free neutrons. To derive the EoS, we first note that the phase space density of particles is

$$\frac{dN}{d^3x d^3p} = \frac{gf}{h^3}, \tag{2.6}$$

where g is the statistical weight, $g = 2S + 1$ for massive particles, with S the spin angular momentum, $g = 2$ for photons and $g = 1$ for neutrinos; $f = f(\mathbf{x}, \mathbf{p}, t)$ is the distribution function which gives the average occupation of a volume cell in phase space; h^3 is the volume of such a phase space cell and h is Planck's constant. The particles of interest here are electrons and neutrons, both fermions with $S = 1/2$. If a configuration of particles are incident upon a surface dA, with unit normal $\hat{\mathbf{n}}$, velocity $\mathbf{v}$, momentum $\mathbf{p}$, the pressure on the surface is the momentum flux (an energy density) given by:

$$P = \int_{2\pi} (\mathbf{p} \cdot \hat{\mathbf{n}})(\mathbf{v} \cdot \hat{\mathbf{n}}) \left(\frac{dN}{d^3x}\right) \frac{d\Omega}{2\pi}. \tag{2.7}$$

The density is simply given by $\rho_{\mathrm{e}} = \mu_{\mathrm{e}} m_{\mathrm{u}} n_{\mathrm{e}}$ for electrons, and $\rho_{\mathrm{n}} = m_{\mathrm{n}} n_{\mathrm{n}}$ for neutrons where

$$n_{\mathrm{e/n}} = \int \left(\frac{dN}{d^3x}\right). \tag{2.8}$$

Using Eq. 2.6, both the P and ρ integrals can be converted to integrals over momentum. If we consider the idealised case of a completely degenerate fermionic gas at $T = 0$ K we know that $f(E) = 1$ below the 'Fermi energy', E_{F}, and is zero above it. Thus we need only integrate up to the 'Fermi momentum' defined by $E_{\mathrm{F}}^2 = p_{\mathrm{F}}^2 c^2 + m_{\mathrm{e}}^2 c^4$. The general solutions for P and ρ are

$$P_{\mathrm{e/n}} = A_{\mathrm{e/n}} \phi(x_{\mathrm{e/n}}), \tag{2.9}$$

$$\rho_{\mathrm{e/n}} = B_{\mathrm{e/n}} x_{\mathrm{e/n}}^3. \tag{2.10}$$

where:

$$\phi(x) = \frac{1}{8\pi^2} \left[x(1 + x^2)^{\frac{1}{2}} \left(\frac{2}{3}x^2 - 1\right) + \ln\left(x + (1 + x^2)^{\frac{1}{2}}\right) \right], \tag{2.11}$$

and the 'e' and 'n' subscripts denote electrons and neutrons, $A_{\mathrm{e/n}}$ and $B_{\mathrm{e/n}}$ are constants listed in Appendix B and $x = p_{\mathrm{F}}/m_{\mathrm{e/n}}c$ is the dimensionless Fermi momentum. These expressions for pressure and density become the familiar polytropic laws for degenerate gases in two extreme scenarios.

- In the non-relativistic limit we have $x \ll 1$, i.e. $\rho_{\mathrm{e}} \ll 10^9$ kg m^{-3}. In this regime we see that $\phi(x) \to x^5/15\pi^2$.

- In the relativistic limit we have $x \gg 1$, i.e. $\rho_e \gg 10^9$ kg m^{-3}. In this regime we see that $\phi(x) \to x^4/12\pi^2$

In both cases we get a polytropic EoS of the form $P = K\rho^\gamma$ where $\gamma_{\text{non-rel}} = 5/3$ and $\gamma_{\text{rel}} = 4/3$. For a degenerate neutron gas, similarly we get two polytropes:

- The non-relativistic case is when $x \ll 1$, i.e. $\rho_n \ll 6.1 \times 10^{18}$ kg m^{-3};
- The relativistic case is when $x \gg 1$, i.e. $\rho_n \gg 6.1 \times 10^{18}$ kg m^{-3}

but note that in the highly-relativistic case the density is dominated by ε_n/c^2.

Low Density Corrections

We can see that in the outer layers, it is the electron pressure which dominates. However, near the stellar surface the density is somewhat lower and the assumption of a degenerate gas breaks down with a smooth transition to planet-like densities [50]. We can account for this by determining a 'Coulomb correction' for our expression for pressure. The most important correction is due to the repulsion of electrons by protons in nuclei. The attraction between electrons is not important as they are distributed much more sparsely. For densities $\gtrsim 10^7$ kg m^{-3} we make the 'Wigner–Seitz approximation', i.e. as $T \to 0$ the ions are in a lattice which maximises the inter-ion separation. Considering spherical shells of radius a, containing an ion of charge Ze and uniformly distributed electrons we can obtain the Coulomb energy per electron which gives a relationship very close to[6] that of a body-centred cubic lattice $E/Z = -1.45079 \times Z^{2/3} e^2 n_e^{1/3}$. The corresponding pressure $n_e^2 d(E/Z)/dn_e$ corrects (lowers) the pressure obtained from Eq. 2.9. For densities $\lesssim 10^7$ kgm^{-3} the non-uniformities in n_e mean alternative corrections are needed (e.g. the Fermi–Thomas method, see Shapiro and Teukolsky [60] and Padmanabhan [50]).

Inverse β Decay

When the density/pressure is sufficiently high such that electrons have energies $E \geq (m_n - m_p)c^2 = 1.29$ MeV inverse β decay (aka 'electron capture') occurs.

$$e^- + p^+ \to n + \nu_e \tag{2.12}$$

Usually this process is balanced by β decay

$$n \to p^+ + e^- + \bar{\nu}_e, \tag{2.13}$$

but above a critical density value, ρ_β, β decay is blocked and electron capture proceeds unbalanced. This is because the density of relativistic degenerate electrons is high, i.e. E_F, which depends on density, is high. As the electrons created in

---

[6] A body-centred cubic lattice has a coefficient of 1.44423.

β decay would have to occupy high energy levels it is more favourable for β decay not to occur. From this point onwards relativistic degenerate electrons penetrate (initially) iron nuclei and form increasingly iron-rich nuclei. This process is known as 'neutronisation'. The density where inverse β decay begins to occur unbalanced, can be calculated (see e.g. Shapiro and Teukolsky [60] or Padmanabhan [50]) to be $\rho_\beta \approx nm_\mathrm{p} \approx 1.2 \times 10^{10}\,\mathrm{kg\,m^{-3}}$.

Neutron Drip

In terrestrial (low-density) scenarios the equilibrium nucleus for a system of A baryons is ^{56}Fe if $A \lesssim 90$. For higher values it is more than one nucleus, but with maximum stability when A is a multiple of 56. So, as A becomes very large, say $\sim 10^{57} \sim 1\,\mathrm{M}_\odot$, the minimum energy composition is pure ^{56}Fe. However at high densities, above ρ_β, this balance shifts as we have a changing n/p ratio in nuclei. With more neutrons the strong force plays a more important role than the repulsive Coulomb forces and the equilibrium nucleus shifts to more neutron-rich values. Neutrons can only be added up to a certain density, $\rho_\mathrm{drip} = 4.3 \times 10^{14}\,\mathrm{kg\,m^{-3}}$, the neutron drip point. For $\rho > \rho_\mathrm{drip}$ neutrons created in inverse β decay drip free from the nucleus. The energy of newly created neutrons can be thought of as their rest mass minus a decreasing (because of degeneracy) effective binding energy so that when $E_\mathrm{F} = m_\mathrm{n}c^2$ it is more energetically feasible for the neutron to be created outside the nucleus. Thus, for $\rho_\mathrm{drip} < \rho < \rho_\mathrm{nuc}$, where $\rho_\mathrm{nuc} = 2.7 \times 10^{17}\,\mathrm{kg\,m^{-3}}$ is nuclear density, we have a two phase structure consisting of a lattice of neutron heavy nuclei and a sea of neutrons. Beyond ρ_nuc all nuclei have dissolved and the system consists mostly of a neutron fluid with a proton fluid and relativistic degenerate electrons.

The equilibrium nucleus for a given $\rho > \rho_\beta$ is determined by minimising the energy density of the system, ε, i.e. finding the nucleus (A, Z) which minimises $\varepsilon = \varepsilon(n, A, Z, Y_\mathrm{n})$. The difficulty lies in specifying the correct form for this energy density, in particular the form of the nucleus binding energy. This essentially requires a choice for which theory and/or approximations to make regarding physics at high densities. The various models predict very similar equilibrium nuclei between ρ_β and ρ_drip and neutron drip is usually determined to occur when the equilibrium nucleus has transitioned to either ^{116}Se, ^{118}Kr or ^{120}Sr (see Xu et al. [74], for recent calculations). Above ρ_nuc the calculations become uncertain as the densities are far above anything reproduceable in a laboratory and physics in this regime is poorly understood.

Superfluids and Magnetic Fields

The above description, although involved, neglects a few salient effects important when considering the neutron star interior—namely rotation and magnetic fields. Considering first rotation we note that the neutron fluid which exists at densities above ρ_drip is a superfluid. It thus possesses a phase ϕ with velocity $v = (\hbar/2m)\nabla\phi$ and can be described by one coherent wave function $\psi \propto n^2 e^{i\phi}$ [50]. Rigid body rotation is described by $\mathbf{v} = \boldsymbol{\Omega} \times \mathbf{r}$ so that $\nabla \times \mathbf{v} = 2\boldsymbol{\Omega}$ but obviously this cannot happen in superfluids. Instead the superfluid breaks into vortices with singular velocities, i.e. $\nabla \times \mathbf{v}$ vanishes everywhere but at the origin of the vortex. With such a singularity the phase of the wave function can change by $2\pi N$, so that the circulation for a single vortex is $\oint \mathbf{v} \cdot d\mathbf{l} = NK$ where $K = h/2m_{\rm n} = 2 \times 10^{-7}\ \mathrm{m^2\,s^{-1}}$, or equivalently that $\nabla \times \mathbf{v} = NK\delta^{(2)}(\mathbf{r})$. For many vortices, say within a radius $\mathbf{r}$, the circulation is $\oint \mathbf{v}(\mathbf{r}) \cdot d\mathbf{l} = K \int_0^r n(r')2\pi r' dr' = KN(r)$, where $n(r)$ is the number of vortices per unit area, and $N(r)$ is the number of vortices enclosed by a circle of radius r. If the vortices are in the z-direction $\oint \mathbf{v} \cdot d\mathbf{l} = 2\pi r v(r) = 2\pi r^2 \Omega(r)$ so we find that

$$\frac{1}{r}\frac{\partial}{\partial r}[r^2\Omega(r)] = Kn(r). \tag{2.14}$$

For rigid body rotation we have no shear ($\partial\Omega/\partial r = 0$) so for a superfluid rotating with $\Omega = 2\pi/P$ a constant density of vortices:

$$n = \frac{2\Omega}{K} = 10^7 \Omega\ \mathrm{m^{-2}} = 2\pi \times 10^7 P^{-1}\mathrm{m^{-2}}, \tag{2.15}$$

is set up with an average vortex spacing of $\approx n^{-1/2} = 120\ \mu\mathrm{m}(P/\mathrm{s})^{1/2}$ [57]. The faster the rotation, the more vortices and the more closely they are packed, and it is these vortices which rigidly rotate with the NS crust. When considering how such a superfluid would slow down we invoke a conservation equation for vortices, i.e. $\partial n/\partial t + \nabla \cdot (n\mathbf{v}_{\rm r}) = 0$ where $\mathbf{v}_{\rm r}$ is a radial vortex velocity [3]:

$$\frac{\partial\Omega}{\partial t} = -\left(2\Omega + r\frac{\partial\Omega}{\partial r}\right)\left(\frac{v_{\rm r}}{r}\right), \tag{2.16}$$

i.e. if there is no radial motion of vortices the superfluid will not slow down. In the "inner crust" (between $\rho_{\rm drip}$ and $\rho_{\rm nuc}$) where we have neutron heavy nuclei in a neutron fluid, the vortices 'pin' to nuclei. A 'pinning region' exists at $\rho_{\rm pin} \sim 7 \times 10^{16}\ \mathrm{kg\,m^{-3}}$ as here the energy cost per particle in creating the normal core of a vortex is reduced by passing through nuclei [2]. Thus vortices are prevented from moving radially outwards, which would enable the superfluid to slow down, by the force associated with the pinning energy. As the NS crust slows down, a lag $\delta\mathbf{v}$ builds up, between the superfluid and crust, resulting in a Magnus force per unit length of the form:

$$F = \rho(\nabla \times \mathbf{v}_{\rm line}) \times \delta\mathbf{v}. \tag{2.17}$$

The pinning force has a maximum value, so that, once surpassed, the vortex lines suddenly unpin, move outwards and transfer angular momentum to the crust. This produces a sudden spin-up of the star that is observed as a 'glitch' in pulsar timing analyses [40], discussed further in Sect. 5.2.1.

Next we consider the effects of magnetic fields. Simply invoking magnetic flux conservation before and after the supernova wherein the NS is formed, we must have very strong magnetic fields in neutron stars, with 10^{12} G being typical (see Sect. 2.3.1). The result of this is that the nuclei in the crust are not spherical but in fact are stretched along magnetic field lines. The stretching can be quite dramatic and the nuclei are sometimes referred to as 'spaghettified'.[7] The proton fluid is a Type II superconductor and forms vortices, each carrying magnetic flux quantised in units of $hc/2e = 2 \times 10^{-11}$ G m^2. Although there is much less proton fluid than neutron fluid, there are many more flux vortices than superfluid vortices, and they are much more densely packed. As the initial magnetic field in a NS is expected to have non-uniform poloidal components, and the flux vortices not subject to a pinning force, the configuration is much more complicated than that of the superfluid vortices [57].

The Core

The composition of the core is unknown, but it is dictated by whether the EoS is "soft" (highly compressible) or "hard" (less compressible). A soft EoS predicts a higher ρ_c, lower maximum mass and lower radii than a hard EoS. If the density is sufficiently high such that $\mu_n - \mu_p = \mu_e > m_\pi c^2 = 139.57$ MeV then we will get pions being created via $n \rightarrow p^+ + \pi^-$. At ρ_{nuc}, $\mu_e \sim 100$ MeV so we might expect pions to be created at $\rho \gtrsim 2\rho_{nuc}$. This would be significant as pions are spin-0 bosons and thus can form a Bose–Einstein condensate where many of the particles are in the lowest energy state. These bosons would have no kinetic energy and so not contribute to the pressure but still contribute to the density (i.e. soft EoS). Another possibility might be strange quark matter.[8] Nucleons begin to 'touch' when the nucleon separation equals the nucleon radius. This corresponds to a density of $\rho_{quark} \sim m_n/((4\pi/3)(\mathrm{fm})^3) \sim$ few $\times \rho_{nuc}$. Beyond this density we might suppose that quarks drip out of nuclei to form quark matter—a degenerate Fermi liquid. Such soft EoS have low maximum masses of $\sim$1.5 $M_\odot$ and maximum radii of $\sim$8 km. A NS with a hard EoS has a much lower central density so that pion condensates and quark matter will not form in the core and maximum masses can be as high as $\sim$3 $M_\odot$. The majority of mass measurements (see Fig. 2.1) are $\sim$1.4 $M_\odot$, favouring a softer EoS, but there are a wide range of observed masses and, for now, many EoS remain consistent with observations.

Figure 2.3 shows a schematic representation of the interior layers of a neutron star from the crust to the core, summarising what has been discussed in Sect. 2.2.2.

---

[7] Other pasta-based vocabulary is also used and this magnetic spaghettification should not be confused with gravitational spaghettification!

[8] This quark matter would consist of up, down and strange quarks. The other three flavours are too massive to be created within NSs.

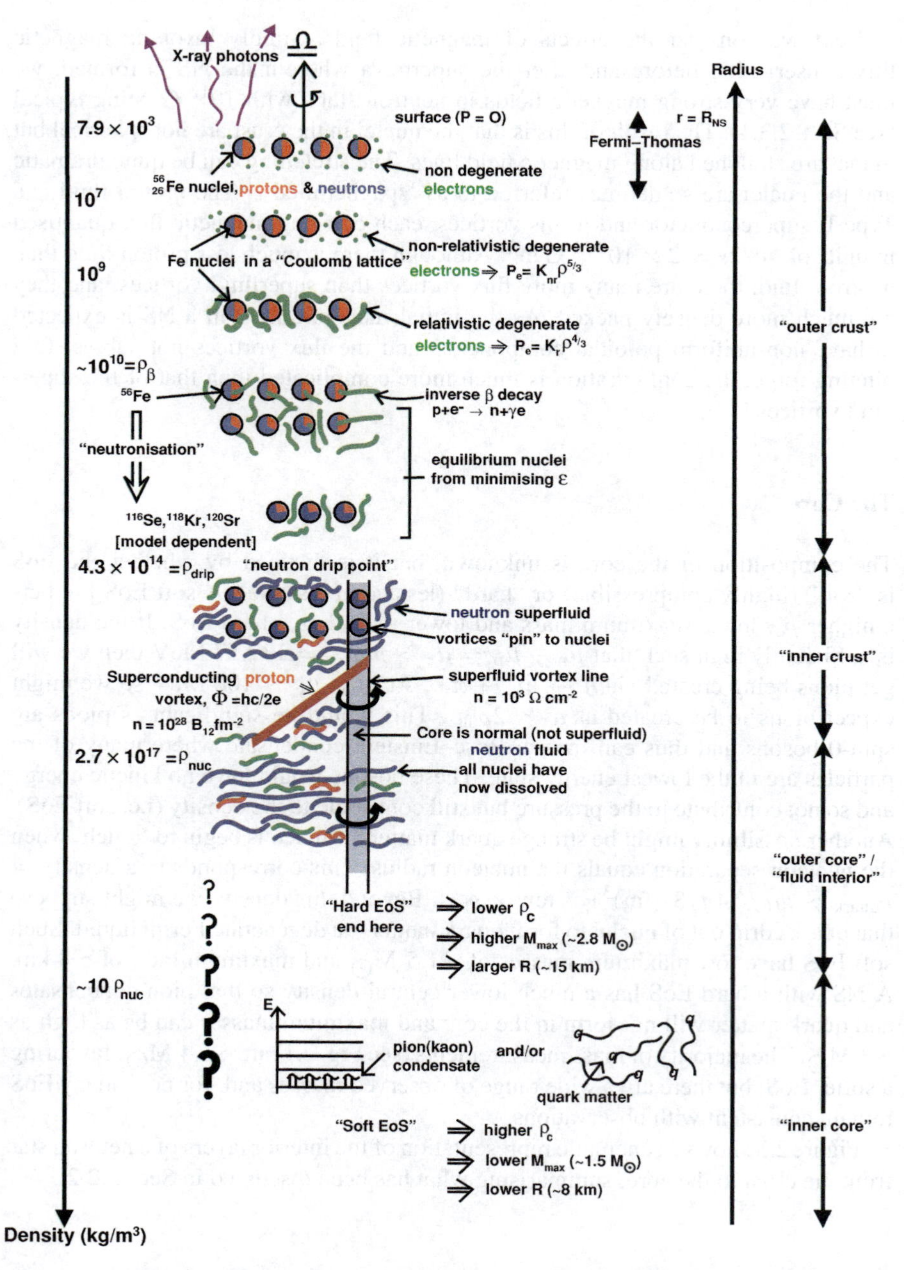

Fig. 2.3 A schematic showing the interior layers of a neutron star from the crust to the interior to the unknown composition of the core, as described throughout Sect. 2.2.2

2.3 Observed Manifestations of Neutron Stars

Although predicted in 1934 by Baade and Zwicky [4], just 2 years after Chadwick discovered the neutron, the concept of a neutron star remained primarily a hypothetical object, existing only in the minds of theorists, until the 1960s. With the discovery of pulsars in 1967 [21], neutron stars were quickly elevated from theoretical curiosities to the cutting-edge of astrophysical research, where they remain to this day. At the time of writing there are approximately 2000 known neutron stars, most of which manifest themselves as pulsars. Although primarily studied at radio wavelengths neutron stars are observable across the entire electromagnetic spectrum, and they are regularly studied in the optical, infrared, X-ray and in γ-rays.

2.3.1 Pulsars

Pacini [49], just before, and Gold [16], just after their discovery, were the first to suggest that pulsars were rotating neutron stars, giving birth to the so-called 'lighthouse model' for pulsar emission. Simply put, the NS is considered to be a rapidly-rotating, highly-magnetised, dense ball, something which is known in laboratories as a terrella. While an Earthly terrella can be built and studied, the results of such an experiment in no way apply to their cosmic counterparts as the magnetic fields involved are many orders of magnitude higher. Without discussing for now the pulsar emission mechanism, but simply considering a rotating highly-magnetised terrella slowing down, we can derive some commonly-used equations of pulsar astronomy. The assumptions made are: (1) The magnetic field in the NS is pure dipolar; (2) Pulsars are powered by the loss of rotational kinetic energy [37, 40]. This loss is:

$$\dot{E}_{rot} = \frac{d}{dt}\left(\frac{1}{2}I\Omega^2\right) = 4\pi^2 I\dot{P}P^{-3}, \tag{2.18}$$

where $P = 2\pi/\Omega$ and we take $I = \frac{2}{5}MR^2$ (uniform sphere). The loss of energy from a rotating magnetic dipole whose magnetic axis is offset from its rotation axis by an angle α (see e.g. Jackson [23]) is:

$$\dot{E}_{\text{dipole}} = \frac{2}{3c^3}|\mu|^2\Omega^4\sin^2\alpha, \tag{2.19}$$

where μ is the magnetic dipole moment. Equating these and inserting the canonical values for R and M to get $I = 10^{38}$ kg m^2 we get:

$$B = 3.2\times10^{19}\ \text{G}\sqrt{\frac{P\dot{P}}{\sin^2\alpha}}, \tag{2.20}$$

where α is the angle between the magnetic and rotation axes and we have used $\mu = B/R^3$. If we assume an orthogonal rotator ($\alpha = 90°$) we get an estimate for the minimum magnetic field strength:

$$B_{\rm min} = 3.2 \times 10^{19}\,{\rm G}\sqrt{P\dot{P}}\,, \tag{2.21}$$

or twice this value for the magnetic field at the poles. If α is randomly distributed then the estimate for an average pulsar ($\alpha = 60°$) would be another factor of 2 higher. We can also determine an age estimate for the pulsar. If the frequency is $\nu = 2\pi\Omega = P^{-1}$ then we can write a 'spin-down law' of the form:

$$\dot{\nu} = -K\nu^n \tag{2.22}$$

where n is known as the 'braking index' and K is usually assumed to be constant. In that case, taking the derivative of this equation and re-arranging we can see that n can be expressed as:

$$n = \frac{\nu\ddot{\nu}}{\dot{\nu}^2}. \tag{2.23}$$

Integrating Eq. 2.22 and using Eq. 2.23 gives the age:

$$T = \frac{-\nu}{\dot{\nu}(n-1)}\left[1 - \left(\frac{\nu}{\nu_{\rm birth}}\right)^{n-1}\right]. \tag{2.24}$$

As we do not, in general, know the birth spin frequency of a pulsar, we cannot evaluate this but we can get an estimate of the age from determining the 'characteristic age'. To do this we assume that: (1) $\nu_{\rm birth} \gg \nu_{\rm now}$ (i.e. the pulsar was born spinning much faster than it is now) and that (2) $n = 3$ (i.e. that we have a purely dipole field) so that

$$\tau = \frac{P}{2\dot{P}} = -\frac{\nu}{2\dot{\nu}}. \tag{2.25}$$

Characteristic ages are commonly quoted but we note that it only equals the true age if our above assumptions hold, although clearly it gives a representative evolutionary timescale for a pulsar with period P and period derivative $\dot{P}$.

Pulsar Magnetospheres on the Back of an (Over-Sized) Envelope

As we have said, the pulsar emission mechanism is not determined. What we might first want to know is the plasma distribution in the pulsar magnetosphere. Then we could, for instance, identify regions of emission, with regions where the magnetospheric currents flow. We consider a rotating conducting terrella in a vacuum. Ohm's law is $\mathbf{J} = \sigma\mathbf{E}'$ where $\mathbf{E}'$ is the electric field in the rotating frame. The Lorentz transform between the stationery frame and the rotating (primed) frame is well known to be $\mathbf{E}' = \gamma(\mathbf{E} + \mathbf{v}\times\mathbf{B}) - (\gamma^2/(\gamma+1))\mathbf{v}(\mathbf{E}\cdot\mathbf{v})$ [23]. Considering non-relativistic rotation speeds the second term drops out,[9] so that for a "highly conducting" sphere we get:

[9] See Lynden-Bell [39] for calculations involving relativistic rotation speeds.

$$\mathbf{E}_{\text{in}} = -(\Omega \times \mathbf{r}) \times \mathbf{B}_{\text{in}}, \tag{2.26}$$

which is known as the 'force-free' or 'MHD' condition. This applies inside the star. We can choose a magnetic field dependence, e.g. a point dipole $\mathbf{B} = B_0(a/r)^3$ $(2\cos\theta, \sin\theta, 0)$ (inside and outside), where B_0 is the magnetic field strength at the equator, as estimated by Eq. 2.21. This is an aligned rotator, chosen for simplicity, in order to give a brief outline. Evaluating Eq. 2.26 then gives the electric field inside the star, and Gauss's Law gives the volume charge density. These turn out to be:

$$\mathbf{E}_{\text{in}} = \frac{\phi_0 a}{r^2}(\sin^2\theta, -2\sin\theta\cos\theta, 0), \tag{2.27}$$

$$\rho_{\text{in}}/\epsilon_0 = -\frac{2\phi_0 a}{r^3} P_2(\cos\theta), \tag{2.28}$$

where $\phi_0 = \Omega B_0 a^2$ is a voltage and $P_2(\cos\theta)$ is the second Legendre polynomial. We can determine the interior electric potential from $\mathbf{E} = -\nabla\Phi$ to be $\Phi = \phi_0 a \sin^2\theta/r$ but we re-write this (for later convenience) as a sum of two Legendre polynomials[10]:

$$\Phi_{\text{in}} = -\frac{2}{3}\phi_0\left(\frac{a}{r} - \frac{a}{r}P_2(\cos\theta)\right). \tag{2.29}$$

As we have a surrounding vacuum, $\rho_{\text{out}} = 0$ by construction, then $\nabla^2\Phi_{\text{out}} = 0$. The general solution to this (Laplace's equation) is $\Phi_{\text{out}} = \sum_{l=0}(a_l/r^{l+1})P_l(\cos\theta)$. Using $\Phi_{\text{in}}(r = a)$ as a boundary condition we can determine the a_l coefficients for Φ_{out} to get:

$$\Phi_{\text{out}} = -\frac{2}{3}\phi_0\left(\frac{a}{r} - \frac{a^3}{r^3}P_2(\cos\theta)\right). \tag{2.30}$$

From this we can calculate the outside electric field and, from the discontinuity of E_r, the surface charge density. These are:

$$\mathbf{E}_{\text{out}} = \frac{\phi_0 a}{r^2}\left(\frac{2}{3r^2} - \frac{2a^2}{r^4}P_2(\cos\theta), -\frac{2a^2}{r^4}\sin\theta\cos\theta, 0\right), \tag{2.31}$$

$$\sigma/\epsilon_0 = -\frac{4\phi_0}{3a^3}P_2(\cos\theta). \tag{2.32}$$

Finally we can calculate $\mathbf{E}\cdot\mathbf{B}$, which gives the magnitude of the electric field component parallel to $\mathbf{B}$. Clearly, by Eq. 2.26 this vanishes within the star. Outside the star it is:

---

[10] $P_0(x) = 1$, $P_1(x) = x$, $P_2(x) = \frac{1}{2}(3x^2 - 1)$.

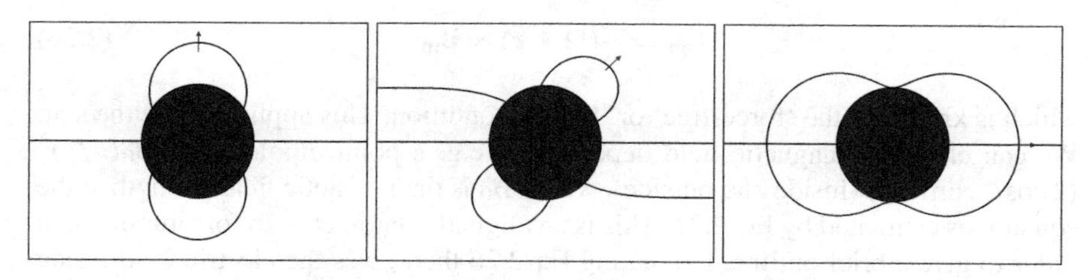

Fig. 2.4 $\mathbf{E} \cdot \mathbf{B} = 0$ surfaces for the vacuum fields, with $\alpha = 0°, 45°$ and $90°$. Particles released from the surface, or indeed placed anywhere within the light cylinder will form a stable dome and torus, sitting on these surfaces. In the case of an orthogonal rotator a 'quad-lobe' structure develops. The arrow denotes the direction of **B**

$$\mathbf{E} \cdot \mathbf{B} = \phi_0 B_0 \frac{a^4}{r^5} \cos\theta \left(\frac{1}{3} - \frac{a^2}{r^2} \cos^2\theta \right), \tag{2.33}$$

which clearly vanishes at $\cos\theta = 0$ (i.e. in the equatorial plane) and at $3\cos^2\theta = (r/a)^2$. These $\mathbf{E} \cdot \mathbf{B} = 0$ surfaces are shown in Fig. 2.4 for this aligned case, and although we have not given the results here (see Appendix B), for two inclined cases also. We have described the 'vacuum fields' for the aligned case but the general fields at any angle, as first determined by Deutsch [14], could likewise be analysed. With these initial conditions, allowing the system to evolve we find that the particles which make up the surface charge will be ripped off. To see that the charges have a work function of effectively zero we note that $\phi_0 = 6 \times 10^{16}$ V. Numerous three-dimensional plasma simulations of this scenario [29, 42, 64] have shown that the particles move away from the surface and form a 'dome-torus' configuration, i.e. the negative charges sit on top of the poles and the positive charges form an equatorial torus, essentially occupying the regions around the $\mathbf{E} \cdot \mathbf{B} = 0$ surfaces.

Now we have reached an impasse, the stable dome-torus solution does not produce a pulsar! The imaginary volume defined by a corotation velocity of c is usually called the 'light cylinder' and has a radius of $R_{\mathrm{LC}} \approx 5R_{\mathrm{NS}}(P/1\ \mathrm{ms})$. Clearly most of light cylinder is empty and charges are not being accelerated. To create emission this apparently stable solution must be altered somehow. This leads us to a description of a famous caricature of a pulsar magnetosphere due to Goldreich and Julian [17]. This model assumes that Eq. 2.28 applies everywhere outside the star so that we get a charge-separated magnetosphere. It also extends the $\mathbf{E} \cdot \mathbf{B} = 0$ condition outside of the star, although this only holds in the closed region, bounded by the last field line which closes within R_{LC}. In the open region $\mathbf{E} \cdot \mathbf{B} \neq 0$, i.e. there is a parallel electric field accelerating particles outwards in the open region. This accelerating potential above the polar caps is the presumed source of the particles which will ultimately produce coherent radiation. There are several terminal problems—there is no current closure and field lines crossing the "null surface" (where $P_2(\cos\theta) = 0$ and the charge density changes sign) show outflowing positive particles on lines firmly rooted in negative charge areas. Furthermore the scenario is unstable, as numerous authors have shown (most recently McDonald and Shearer [42], but see references therein) that a Goldreich–Julian charge distribution is unstable and will always collapse into the stable dome-torus configuration.

To overcome this 'dead end', most recent work has simply assumed a magnetosphere with abundant plasma so that $\mathbf{E} \cdot \mathbf{B} = 0$ 'everywhere'. Such a 'force-free magnetosphere' obeys $\rho\mathbf{E} = -\mathbf{J} \times \mathbf{B}$. If the enclosed magnetic flux is Ψ then this can be re-written as:

$$(1 - x^2)\left(\Psi_{,xx} + \frac{1}{x}\Psi_{,x} + \Psi_{,zz}\right) - 2x\Psi_{,x} = -R_{\rm LC}^2 A\frac{dA}{d\Psi}, \tag{2.34}$$

which is known as the 'pulsar equation' [46], where $x = r/R_{\rm LC}$ and the poloidal current I is given by $(\mu_0/2\pi)I = A(\psi) = B_\phi r$. This equation is non-linear and only a few special-case solutions were initially determined, e.g. a split-monopole with non-zero poloidal current and a dipole solution with zero poloidal current which only applies within the light cylinder [47]. The various solutions are unsatisfactory and suffer from being either non-dipolar, not having current closure or kinks/discontinuities at the light cylinder. In 1999, Contopoulos et al. [10] presented the apparently unique solution to the pulsar equation with continuous Ψ across the light cylinder boundary. The solution has a dipolar region closed within the light cylinder and an open region which asymptotically monopolar field lines. The null surface has no special significance as it is just the net charge density which changes sign in the high density plasma [63]. There is an outflowing poloidal current in the open field line region (where the field has a toroidal component) and a return current sheet along the open-closed boundary and the open equatorial plane. The last closed field line on the equatorial plane forms a Y-point which need not exactly be at the light cylinder [67]. Spitkovsky [62] has confirmed this solution and added time-dependence at all inclination angles. This has enabled the determination of the spin-down energy loss rate of a pulsar with such a force-free magnetosphere, and consequently an estimate of the magnetic field strength. These are:

$$\dot{E} = \frac{1}{c^3}|\mu|^2\Omega^4(1 + \sin^2\alpha), \tag{2.35}$$

$$B = 2.6 \times 10^{19}\ \mathrm{G}\sqrt{\frac{P\dot{P}}{1 + \sin^2\alpha}}, \tag{2.36}$$

which we can compare with Eqs. 2.19 and 2.21 determined for vacuum fields. This predicts that misaligned rotators spin down quicker and thus that there would be an excess of aligned rotators at the death line for which there is some evidence [11]. Weltevrede and Johnston [71] have also noted that the number of pulsars with interpulses (thought to be orthogonal rotators) is too low to be explained by a random α distribution, although the most recent simulations [56] show no improvement in reproducing the observed pulsar population using the Contopoulos and Spitkovsky prescription over the standard dipole spindown. The model can also account for braking indices $n < 3$ if the closed region does not extend to the light cylinder. They also predict extremely large braking indices for pulsars near the death line which may be observable (see Sect. 7.1). Some recent successes have included the modelling of realistic γ-ray light curves under the assumption that the current is a direct measure of the emission regions [5, 6] and the simulations of the famous Crab plerion [26, 27]. The missing link in this picture is why the dome-torus solution might evolve to the force-free configuration. The diocotron instability, due to velocity shear of the orbiting charges, may provide such a link [51, 64] although this is still a matter of ongoing research.

2.3.2 RRATs

In 2006, a group of sources known as Rotating RAdio Transients (RRATs[11]) was discovered, by McLaughlin et al. [44]. This section gives an overview of the understanding of RRATs at the beginning of the research towards this thesis, i.e. in 2007. Chapters 8 and 9 give an up to date and detailed overview of current knowledge, i.e. in late 2010, incorporating the research presented in subsequent chapters and that of various others authors throughout the past few years.

Eleven RRATs were discovered by McLaughlin et al. [44] in an archival search for single dispersed bursts in the Parkes Multi-beam Pulsar Survey (PMPS, see Chap. 4 for a detailed description of the survey). These sources are characterised by detectable single pulses at frequencies of 1.4 GHz, with peak flux densities of 0.1–3.6 Jy. The pulse widths observed range from 2 to 30 ms and the pulses occur with burst rates of $(3\ \mathrm{h})^{-1}$ to $(4\ \mathrm{min})^{-1}$. The RRATs were not seen as periodic sources in Fourier domain searches and consecutive pulses were not detected. This suggests that RRATs are either much more weakly emitting, or in fact 'off', during times when we do not detect pulses. Despite this, by examining time differences between the arrival times of pulses, underlying periods were identified for all eleven sources. The periods thus detected are 'long', e.g. six of the sources have periods longer than 4 s, as compared to just 0.5% of the radio pulsar population. Monitoring these periods over time using pulsar timing techniques (see Chap. 5) revealed that three of the sources had a measurable slow-down rate. From this knowledge of $\dot{P}$, standard pulsar equations, given in Sect. 2.3.1 can be used to infer values for $\dot{E}$, B and τ. One of the sources, J1819−1458, has an inferred minimum equatorial magnetic field strength of 5×10^{13} G (vacuum) or 3×10^{13} G (force-free), which places it above the 'photon-splitting' line ([7], discussed in Chap. 8). Using Eq. 1.1 and the dispersion measure-derived distances, the brightness temperature of the pulses can be inferred to be 10^{22}–10^{23} K, and we have included the RRATs in Fig. 1.2. The maximum source size, as implied by the pulse widths, and causality, is in the range 600–9000 km. This means that RRATs, whatever they may be, are compact objects emitting coherent non-thermal radiation. The inferred source sizes, the underlying rotation periods, as well as the expected time-scales for transient behaviour all point towards RRATs being Galactic neutron stars.

---

[11] Although I prefer the name eRRATic neutron stars.

X-ray observations of J1819–1458, the most prolific source, using first Chandra [55] and then XMM Newton [45], have detected a thermal spectrum with $kT \sim 140$ eV, characteristic of a cooling neutron star. Pulsations at the period determined from radio were also observed, but no X-ray bursts [58], as well as a 0.5 keV spectral absorption feature, tentatively identified with proton cyclotron resonant scattering. The magnetic field strength inferred from this would be 2×10^{14} G, implying $\alpha = 15°$ for the vacuum estimate,[12] which might imply a wide pulse profile, unlike what is observed. Interestingly, equating the cyclotron estimate with the force-free expression is impossible as it yields $\sin^2 \alpha < 0$ which cannot be satisfied for any real α. Thus, if the force-free model is correct the interpretation of the cyclotron line is incorrect, and vice-versa. An X-ray candidate for another source was put forward by Hoffman et al. [22], for J1911+00, which fortuitously occupies the same field as Aql X-1, a much-observed LMXB. No pulsations or bursts were detected and the spectrum is not thermal. With by far the lowest radio burst rate, J1911+00 is unlikely to ever have an accurate position determined, and hence progress in identifying this X-ray source is unlikely.

One of the most interesting characteristics of RRATs is the size of their inferred population. A population synthesis simulation was performed to determine the total Galactic population of RRATs ([44], D. R. Lorimer, private communication), assuming the observed pulsar luminosity distribution. The simulation gives an estimate for the total number of RRATs (seen + unseen) which is

$$N_{\mathrm{RRAT}} \approx 2 \times 10^5 \left(\frac{100\ \mathrm{mJy\ kpc^2}}{L_{\mathrm{min}}}\right)\left(\frac{0.5}{f_{\mathrm{on}}}\right)\left(\frac{0.5}{1 - f_{\mathrm{RFI}}}\right)\left(\frac{0.1}{f_{\mathrm{beam}}}\right) \tag{2.37}$$

and depends on three parameters: f_{on} is the fraction of sources which had bursts during the 35-min PMPS observations, f_{RFI} is the fraction of these bursts missed due to impulsive radio frequency interference (RFI) and f_{beam} is the fraction of these bursts which were beamed towards Earth. We note that this estimate is also very sensitive to $\mathrm{L_{min}}$, which is very uncertain (D. R. Lorimer, private communication). Furthermore it requires the active Galactic pulsar population as an input. Here it was taken as $\sim 10^5$ [36, 69]. The estimate thus predicts that we have $N_{\mathrm{RRAT}} = \gamma N_{\mathrm{PSR}}$ with $\gamma \approx 1$–4. The discovery paper proposes RRATs as a previously unknown population of neutron stars distinct from radio pulsars [44] whose numbers in the Galaxy are comparable to or even much larger than the long-studied radio pulsars.

Figure 2.5 shows the observed and inferred radio properties of the original eleven RRAT sources as they were known in 2007. These are contrasted with the properties observed in radio pulsars. A few remarks immediately come to mind. The first is that we do not have much information—for any properties inferred from $\dot{P}$ measurements we have information for only 3 RRATs, and at most 11 for the other properties. The high periods of the RRATs are evident although they seem unremarkable in all other senses except for the high $\dot{P}$, and hence high B of J1819–1458. What is noteworthy is that only one of the of the six sources with $P > 4$ s have a determined

[12] Not 30°, as published in McLaughlin et al. [45].

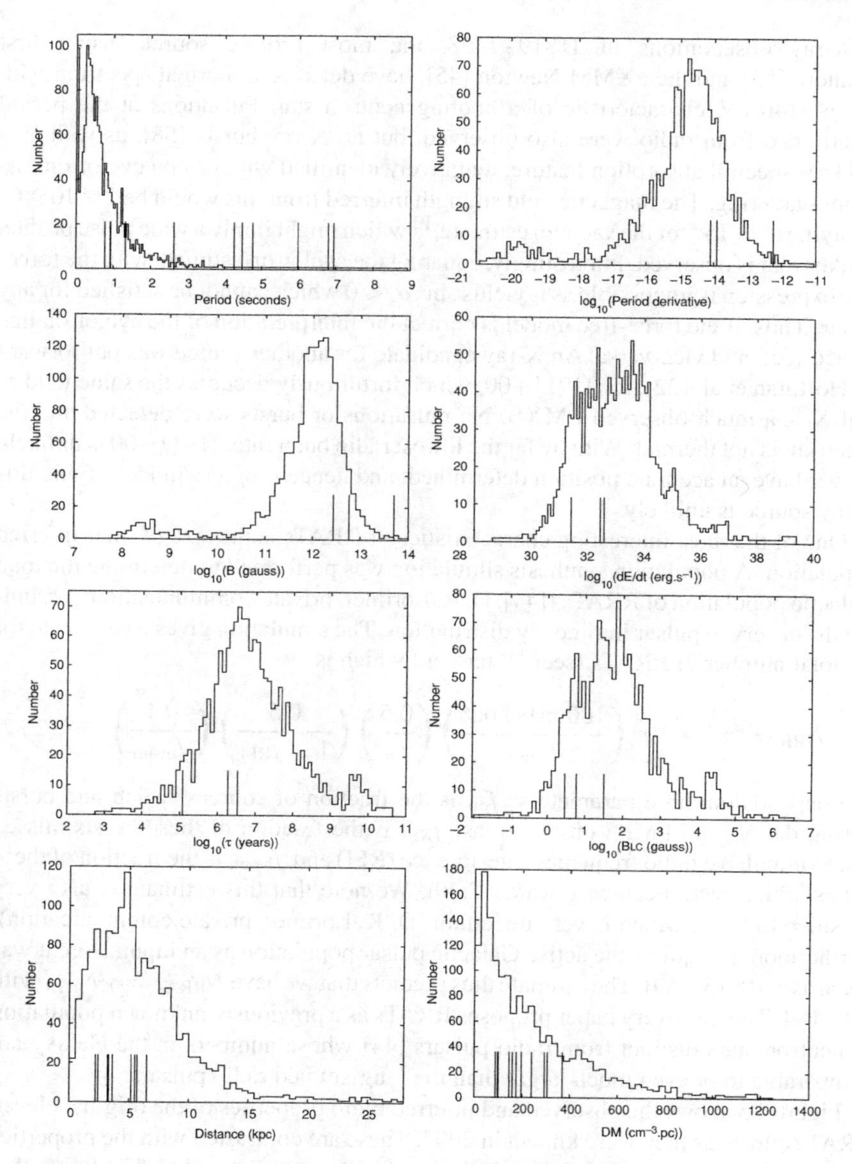

Fig. 2.5 In each plot the distribution is that of the overall radio pulsar population. Overplotted in each case are up to 11 impulses denoting the parameter values of the original PMPS RRATs. The parameters plotted are: P, $\dot{P}$, B, $\dot{E}$, τ, B_{LC}, distance and DM. In some cases the abscissa is plotted linearly, in some as a base-10 logarithm

$\dot{P}$ and none of these can fall into a 'normal' area of $P - \dot{P}$ space. They will either be high B sources (possibly in the magnetar region) or old 'dying' pulsars. All of this, combined with our knowledge of their sporadic burst rates and hence the difficulty in detection, the pulse brightness properties (see e.g. Fig. 1.2) and the X-ray observation of J1819−1458 summarises the known properties as of 2007.

Models and Ideas

The discovery of RRATs has been one factor in a recent renewed interest in radio transients, and there have been a number of papers whose aim is to explain the nature of their sporadic emission. A number of these models assume that the emission of RRATs is in some way special, in the sense that additional explanatory factors are needed, above and beyond whatever is needed for pulsar emission, which, as we have discussed, is itself not understood in detail. The ideas put forward include:

Giant Radio Pulse Emitting Pulsars. Perhaps RRATs are regularly emitting, though weak, radio pulsars, which show occasional so-called 'giant radio pulses' (GRPs). GRPs are formally (and arbitrarily) defined as single pulses with intensities greater than ten times the mean [19]. According to this criterion the single pulses from RRATs are GRPs. GRPs have been observed in at least 14 pulsars—most notably the Crab, three millisecond pulsars B1821−24, B1937+21 and J1823−3021A and the young pulsar B0540−69 (see [25] for a complete list). The GRP mechanism is not understood but a number of common features have been identified amongst the GRP sources. These are: (1) Power-law amplitude distributions, i.e. $dN/dS \propto S^{-\alpha}$ with indices $\alpha = 2$–3; (2) High values of $B_{\rm LC}$, the magnetic field strength at the light cylinder radius; (3) High $\dot{E}$, e.g. the five sources named above all have $\dot{E} > 10^{36}$ erg s^{-1}; (4) Very narrow pulses, as short as ~1 ns in the case of the Crab; (5) Emission confined to a narrow pulse-phase window e.g. the Crab GRPs are always at the same phase of the main pulse or inter-pulse; (6) A high degree of polarisation.

RRATs seem to fail these criteria: their amplitude distributions have a gentler $\alpha = -1$ index [44], as can be seen from Fig. 2.5 they have low $B_{\rm LC}$ (~5–30G as compared to ~10^6 G for the Crab), low $\dot{E}$ (10^{31}–10^{33} erg s^{-1}), and their single pulses seem to be a few milliseconds in width and well-resolved. However, the initial pulse-amplitude distributions contain only a small number of pulses and there is no polarisation information. Also, a wide bandwidth study of RRAT pulses, to investigate, for example, if they could consist of bunches of much narrower pulses (see e.g. Jessner et al. [24] and Popov et al. [53]) has not been performed. Furthermore, GRP emission has been associated with enhanced optical emission [61] so an optical study of some of the RRAT sources should help investigate the GRP hypothesis. Such a study of J1819−1458 is presented in Chap. 7.

Distant versions of PSR B0656+14. There are a group of three pulsars known as the 'Three Musketeers'—PSR B0656+14, B1055−52 and Geminga, all of which exhibit high-energy emission. Of interest is B0656+14, a nearby source ($D = 288$ pc) which exhibits brief, but powerful bursts of radio emission which can reach up to 116 times the mean pulse energy [72, 73]. Despite this, the pulses are not considered 'classical' GRPs as they are not sufficiently narrow, nor are they confined to narrow regions of pulse phase and the amplitude distribution is log-normal (i.e. these pulses are much rarer than in the GRP sources). Nonetheless,

B0656+14 constitutes another example of a pulsar with a high modulation index.[13] It has been shown by Weltevrede et al. [72, 73] that if it were (say) 12 times further away (a typical RRAT distance) only ∼1 pulse per hour would be detected from B0656+14 and it would then be detected as a RRAT. This suggests that some or all of the RRATs may be pulsars with extreme pulse amplitude modulation like B0656+14 and weak underlying emission. RRATs were discovered at 1.4 GHz, a relatively high frequency, so that, if they were indeed distant Musketeers, then more pulses would be expected at lower frequencies, where pulsars are usually brighter.

External Trigger Mechanisms. Several authors have proposed external triggers for producing a sudden activation of pulsar-like emission as explanations for the observed RRAT properties. These include models where neutron stars have a surrounding fall-back (or ISM-captured) disc which prevents emission by plugging any acceleration 'gaps' in the magnetosphere. The density instabilities in the disc would provide windows wherein emission could occur. Li [34] argues that the instability timescales are comparable to RRAT burst times. However it is not clear if sufficiently low-mass debris discs exist, as only one fall-back disc has been observed, which is, in any event, too massive and is a passive disc [70]. It has been suggested by Cordes and Shannon [12] that inactive outer gap regions could be re-activated by a particle flow from surrounding asteroidal material. Such material can more easily penetrate the light cylinder if $R_{\rm LC}$ is large, i.e. for long-period pulsars. The model predicts timescales from seconds to weeks which might explain the nulling behaviour seen in some pulsars (see Chaps. 5 and 8), as well as episodic asteroid migration which might explain quasi-periodicities. At present it is difficult to see how to observationally test the model as the asteroidal material is proposed to be sufficiently low mass ($\sim 10^{-6}$ $M_{\odot}$) that it would not yield a detectable imprint in pulsar timing observations (such as a planet). Another proposed mechanism involved the disturbance of radiation belts of plasma (trapped by the magnetic mirror effect), in the co-rotating magnetosphere, by Alfvén waves, caused by starquakes or shear waves in the crust. This could produce particles diffusing downwards towards the star and radiation via the ECM or plasma emission [38]. We note that the discovery of RRATs was contemporaneous with the published results for the 'intermittent pulsar', PSR B1931+24 [28], a pulsar which is 'on' for ∼1 week before switching 'off' for ∼1 month, and many of these external trigger mechanisms are proposed to explain its behaviour also.

Dying Pulsars. It has been suggested that RRATs are pulsars which lie just beyond the death-line [75] with no emission except for transient hot spots where beaming conditions might be turned on temporarily. At first glance, this does not seem to be the case for the original three sources with determined $\dot{P}$ which do not lie near canonical pulsar death-lines [9] and are 'young' by the measure of τ. However we re-iterate that the unmeasured $\dot{P}$ values for the high period RRATs will be in/close to the 'death valley' region of $P - \dot{P}$ space if they do not occupy thehigh B/magnetar

[13] The difference between B0656+14, and, although not mentioned, the 'giant-micropulses' seen in Vela, are quite arbitrary. There is little evidence that there is any physical difference in these situations. We will, at times, refer to all of these phenomena collectively as "giant pulses".

region. We will revisit this scenario in Chaps. 6, 7 and 9 in light of newly detected sources (see Chap. 4).

Miscellaneous. Some models seem more unlikely, such as precession [41] where the beam of a steadily emitting pulsar moves in and out of our line of sight so that we see it only sporadically. This would result in a periodicity which is not observed and so does not explain why the times between observed RRAT pulses are apparently random. It would also need to be fine-tuned to produce single bursts.

Outstanding Questions

As we have outlined above, many explanations have been posited to explain what RRATs might be. What we can say from the initial observations is that they seem to be rotating neutron stars. However many questions remain, in particular we ask *Are RRATs Special*?, or in more specific terms:

(1) Are there really as many RRATs in the Galaxy as the initial estimates imply? How well do we know the parameters in Eq. 2.37?
(2) Are they truly a distinct population? What are the implications of this?
(3) Why do they have longer periods than the radio pulsars? Is this significant?
(4) What decides whether a NS will manifest itself as a RRAT as opposed to (say) a magnetar or an XDINS (Chap. 3) which have the same rotational properties?
(5) Are their observed properties a result of selection effects in our search methods or truly a representation of a class of neutron stars? Given the parameters of our survey and searches, are these the kind of sources we expect to find?
(6) How different is their emission in comparison to the radio pulsar population?
(7) What are their long-term timing properties? How stable, or not, are these properties?
(8) Are they old or young? Are they evolutionarily linked to any of the previously known classes of neutron star?
(9) Can we characterise the observed properties more completely? Are more timing solutions possible and where in $P - \dot{P}$ space do RRATs really live?
(10) Can we discover new sources and improve the characterisations to help to answer all the above questions and identify any key relationships?

In the following chapters we attempt to address these questions. Having summarised our knowledge as of 2007 we will describe what progress has been made in the last 3 years. First, we take at face value the claim that RRATs are a distinct population of neutron stars and investigate the implications of this for the Galactic population.

References

1. B.P. Abbott et al., ApJ **713**, 671 (2010)
2. M.A. Alpar, ApJ **213**, 527 (1977)
3. M.A. Alpar, P.W. Anderson, D. Pines, J. Shaham, ApJ **276**, 325 (1984)

4. W. Baade, F. Zwicky, Proc. Nat. Acad. Sci. **20**, 259 (1934)
5. X. Bai, A. Spitkovsky, ApJ **715**, 1282 (2010)
6. X. Bai, A. Spitkovsky, ApJ **715**, 1270 (2010)
7. M.G. Baring, A.K. Harding, ApJ **507**, L55 (1998)
8. S. Bogdanov, J.E. Grindlay, C.O. Heinke, F. Camilo, P.C.C. Freire, W. Becker, ApJ **646**, 1104 (2006)
9. K. Chen, M. Ruderman, ApJ **402**, 264 (1993)
10. I. Contopoulos, D. Kazanas, C. Fendt, ApJ **511**, 351 (1999)
11. I. Contopoulos, A. Spitkovsky, ApJ **643**, 1139 (2006)
12. J.M. Cordes, R.M. Shannon, ApJ **682**, 1152 (2008) (astro-ph/0605145)
13. J. Cottam, F. Paerels, M. Mendez, Nature **420**, 51 (2002)
14. A.J. Deutsch, Ann. d'Astrophys. **18**, 1 (1955)
15. N.K. Glendenning, Phys. Rev. D **46**, 4161 (1992)
16. T. Gold, Nature **218**, 731 (1968)
17. P. Goldreich, W.H. Julian, ApJ **157**, 869 (1969)
18. A. Heger, C.L. Fryer, S.E. Woosley, N. Langer, D.H. Hartmann, ApJ **591**, 288 (2003)
19. K.H. Hesse, R. Wielebinski, A&A **31**, 409 (1974)
20. J.W.T. Hessels, S.M. Ransom, I.H. Stairs, P.C.C. Freire, V.M. Kaspi, F. Camilo, Science **311**, 1901 (2006)
21. A. Hewish, S.J. Bell, J.D.H. Pilkington, P.F. Scott, R.A. Collins, Nature **217**, 709 (1968)
22. K. Hoffman, R.E. Rutledge, D.B. Fox, A. Gal-Yam, S. B. Cenko, ArXiv Astrophysics e-prints (2006). astro-ph/0609092
23. J.D. Jackson, *Classical Electrodynamics* (Wiley, 1962)
24. A. Jessner et al., ArXiv e-prints (2010). astro-ph/1008.3992
25. H.S. Knight, Chin. J. Astronomy Astrophys. Suppl. **6**, 41 (2006)
26. S.S. Komissarov, MNRAS **367**, 19 (2006)
27. S.S. Komissarov, Y.E. Lyubarsky, MNRAS **349**, 779 (2004)
28. M. Kramer, A.G. Lyne, J.T. O'Brien, C.A. Jordan, D.R. Lorimer, Science **312**, 549 (2006)
29. J. Krause-Polstorff, F.C. Michel, MNRAS **213**, 43P (1985)
30. P. Kroupa, MNRAS **322**, 231 (2001)
31. P. Kroupa, Science **295**, 82 (2002)
32. J.H. Lattimer, M. Prakash, Science **304**, 536 (2004)
33. J.M. Lattimer, M. Prakash, ApJ **550**, 426 (2001)
34. X.D. Li, ApJ **646**, L139 (2006)
35. D.R. Lorimer, Living Reviews in Relativity **11**, 8 (2008)
36. D.R. Lorimer et al., MNRAS **372**, 777 (2006)
37. D.R. Lorimer, M. Kramer, *Handbook of Pulsar Astronomy* (Cambridge University Press, 2005)
38. Q. Luo, D. Melrose, MNRAS **378**, 1481 (2007)
39. D. Lynden-Bell, Phys. Rev. D **70**, 104021 (2004)
40. A.G. Lyne, F.G. Smith, Pulsar Astronomy. 3rd edn. (Cambridge University Press, Cambridge, 2004)
41. I.F. Malov, ArXiv Astrophysics e-prints (2007). astro-ph/07110502
42. J. McDonald, A. Shearer, ApJ **690**, 13 (2009)
43. C.F. McKee, E.C. Ostriker, Ann. Rev. Astr. Ap. (2007). astro-ph/0707.3514
44. M.A. McLaughlin et al., Nature **439**, 817 (2006)
45. M.A. McLaughlin et al., ApJ **670**, 1307 (2007)
46. F.C. Michel, ApJ **180**, L133 (1973)
47. F.C. Michel, Theory of Neutron Star Magnetospheres. (University of Chicago Press, Chicago, 1991)
48. J.R. Oppenheimer, G. Volkoff, Phys. Rev. **55**, 374 (1939)
49. F. Pacini, Nature **216**, 567 (1967)
50. T. Padmanabhan, Theoretical Astrophysics, Volume 2: Stars and Stellar Systems. (Cambridge University Press, Cambridge, UK, 2001)

51. J. Pétri, A&A **503**, 1 (2009)
52. P. Podsiadlowski, N. Langer, A.J.T. Poelarends, S. Rappaport, A. Heger, E.D. Pfahl, ApJ **612**, 1044 (2004)
53. M. Popov et al., PASJ **61**, 1197 (2009)
54. D. Prialnik, An Introduction to the Theory of Stellar Structure and Evolution. (Cambridge University Press, Cambridge, UK, 2000)
55. S. Reynolds et al., ApJ **639**, L71 (2006)
56. J.P. Ridley, D.R. Lorimer, MNRAS **404**, 1081 (2010)
57. M. Ruderman, in *Astrophysics and Space Science Library*, vol. 357, ed. by W. Becker, ASSL, p. 353 (2009)
58. R.E. Rutledge, ArXiv Astrophysics e-prints (2006). astro-ph/0609200
59. E.E. Salpeter, ApJ **121**, 161 (1955)
60. S.L. Shapiro, S.A. Teukolsky, Black Holes, White Dwarfs and Neutron Stars. The Physics of Compact Objects. (Wiley, New York, 1983)
61. A. Shearer, B. Stappers, P. O'Connor, A. Golden, R. Strom, M. Redfern, O. Ryan, Science **301**, 493 (2003)
62. A. Spitkovsky, ApJ **648**, L51 (2006)
63. A. Spitkovsky, in *American Institute of Physics Conference Series*, vol. 983, ed by C. Bassa, Z. Wang, A. Cumming, V. M. Kaspi , 40 Years of Pulsars: Millisecond Pulsars, Magnetars and More, p. 20 (2008)
64. A. Spitkovsky, J. Arons, in *Neutron Stars in Supernova Remnants Astronomical Society of the Pacific*, ed. by P. O.Slane, B. M.Gaensler (San Francisco, 2002), p. 81
65. W.W. Stahler, F. Palla, The Formation of Stars. (Wiley, Weinheim, Germany, 2004)
66. I.H. Stairs, Science **304**, 547 (2004)
67. A.N. Timokhin, MNRAS **368**, 1055 (2006)
68. E.P.J. van den Heuvel, in *American Institute of Physics Conference Series*, vol. 924, ed. by T. di Salvo, G. L. Israel, L. Piersant, L. Burderi, G. Matt, A. Tornambe, M. T. Menna, The Multicolored Landscape of Compact Objects and Their Explosive Origins, p. 598 (2007)
69. N. Vranesevic et al., ApJ **617**, L139 (2004)
70. Z. Wang, D. Chakrabarty, D.L. Kaplan, Nature **440**, 772 (2006)
71. P. Weltevrede, S. Johnston, MNRAS **387**, 1755 (2008)
72. P. Weltevrede, B.W. Stappers, J.M. Rankin, G.A.E. Wright, ApJ **645**, L149 (2006)
73. P. Weltevrede, G.A.E. Wright, B.W. Stappers, J.M. Rankin, A&A **458**, 269 (2006)
74. J. Xu, L. Chen, B. Li, H. Ma, ApJ **697**, 1549 (2009)
75. B. Zhang, J. Gil, J. Dyks, MNRAS **374**, 1103 (2007)

Chapter 3
On the Birthrates of Galactic Neutron Stars

The following chapter is an enhanced version of a paper published in the Monthly Notices of the Royal Astronomical Society in 2008: Keane and Kramer [27]. As presented, this chapter, presents investigations based upon the state of knowledge as it was in 2008. It is presented in this way so as to keep the logical flow of the thesis chronological. The following chapters will expand on what has been learnt up to the present, and, in particular, Chaps. 8 and 9 give an up to date summary and overview, incorporating both the work in this thesis and that of many other authors.

3.1 Introduction

In the standard scenario, neutron stars (NSs) are formed during the core collapse of massive stars which links their number in the Galaxy to the Galactic supernova rate. The number of Galactic NSs can be inferred from observations, taking the various manifestation of NSs into account. In recent years, new and different observational manifestations of NSs have been discovered, so that it is warranted to study the impact, if any, of these discoveries on the birthrates that are required to sustain this increased NS population.

The new manifestations of NSs include Rotating Radio Transients (RRATs; McLaughlin et al. [38] and see Sect. 2.3.2) and X-ray Dim Isolated Neutron Stars (XDINS; see [22] and references therein). These objects join the ~1800 known radio pulsars, the small group of magnetars [65] and the central compact objects (CCOs; see e.g. Pavlov et al. [42]). Do these previously unknown types of observable NSs increase the overall population by an amount such that it is difficult to reconcile the formation rates with those predicted by theory and measured from observations? The basic requirement we make to answer this question is that the individual birthrates of the different NS populations should not exceed the Galactic core-collapse supernova (CCSN) rate, i.e.

$$\beta_{\rm CCSN} \geq \beta_{\rm total} = \beta_{\rm PSR} + \beta_{\rm XDINS} + \beta_{\rm RRAT} + \beta_{\rm magnetar} + \beta_{\rm CCO}, \tag{3.1}$$

where $\beta_{\rm X}$ is the birthrate (per century) of a NS of type X.

E. F. Keane, *The Transient Radio Sky*, Springer Theses,
DOI: 10.1007/978-3-642-19627-0_3, © Springer-Verlag Berlin Heidelberg 2011

Recently, this question has also been addressed by Popov et al. [48] where it was concluded that this requirement can be met if we assume that XDINSs are in fact nearby RRATs. However, as we detail below, the pulsar birthrate they considered is a lower limit which has since been superseded. In addition, the recent non-detection of any radio RRAT-like bursts from the XDINSs [29] means that the identification of these two populations is not certain. Furthermore recent work suggests that the heretofore neglected magnetar contribution may not be negligible, so that the question, as to whether the CCSN rate requirement is satisfied, is reinstated. Our aim is to study the posed question by investigating the most recent knowledge about each contributing NS population and its Galactic birthrate. After introducing each manifestation of NS in turn, we will revisit the estimates for all terms in Eq. 3.1. The results are then discussed in detail before conclusions are drawn.

3.2 Different Manifestations of Neutron Stars

Radio Pulsars

As we briefly outlined in Sect. 2.3.1, radio pulsars are rapidly-rotating, highly-magnetised NSs. Coherent radio emission is produced by a pair plasma above the magnetic polar caps of the NS, believed to originate from particle cascades after an acceleration of electrons and positrons in the strong electric and magnetic fields (e.g. Lorimer and Kramer [34]). The spectra for this emission typically increases with decreasing radio frequency with mean spectral index of -1.8 [37] before peaking in the range 100–300 MHz [36]. Pulsar periods range from 1.4 ms up to 8.5 s with two distinct distributions—the 'normal' radio pulsars which have periods of ~500 ms and the so-called 'millisecond pulsars' with typical periods of ~5 ms. Figure 3.1 shows a $P-\dot{P}$ diagram, a standard pulsar classification tool, where these two populations are easily identified. The standard model of pulsar physics assumes pulsars have dipolar magnetic fields and that the loss of rotational energy powers the pulsar. As we have shown in Sect. 2.3.1 this enables us to determine expressions (Eqs. 2.18, 2.21 and 2.25) for the characteristic surface magnetic field, the spin-down energy loss rate and the characteristic age—B, $\dot{E}$ and τ. Lines of constant $B, \dot{E}$ and τ are shown in Fig. 3.1 along with different evolutionary paths for different braking indices. The lower right area of the diagram devoid of any pulsars is known as the pulsar 'death valley' [6]. It is here that it is believed that the electric potential at the polar caps is insufficient for ripping particles from the NS surface, hence failing to provide the plasma needed for radio emission.

Millisecond Pulsars and X-ray Binaries

The standard evolutionary picture for millisecond pulsars (e.g. Alpar et al. [1]) is that they are born in supernovae with periods of 10 s of milliseconds, then evolve along

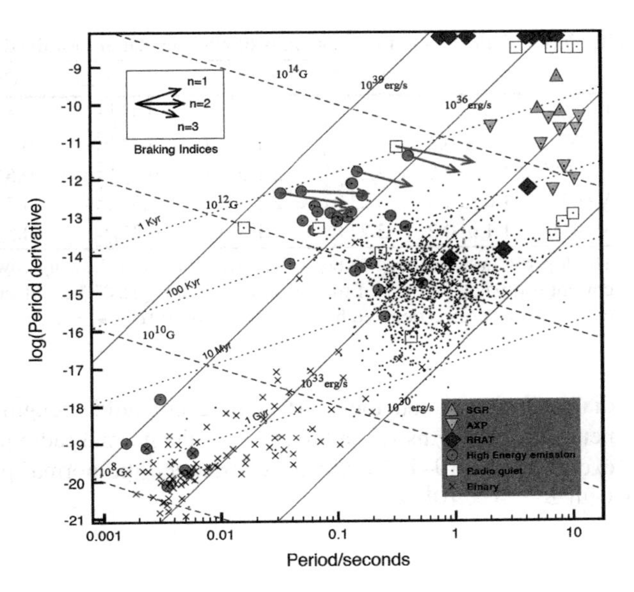

Fig. 3.1 A $P - \dot{P}$ diagram showing the various NS populations. The 4 XDINSs and 8 RRATs without a known $\dot{P}$ are placed at the top right at their respective periods. *Arrows* indicate the current spin evolution of those sources for which a braking index has been measured (Image credit: C. M. Espinoza)

a line of approximately constant magnetic field strength (i.e. $n = 3$) on the $P - \dot{P}$ diagram, slowing down until eventually radio emission ceases once they pass into the pulsar death valley. Here, those 'dead' pulsars which happen to be in binary systems can undergo accretion from their binary companion. This accretion can heat areas of the NS surface ('hot spots') which emit strongly in X-rays—the system is now an X-ray binary [64]. As well as heating the star the accretion can spin up the star to periods of a few milliseconds. The pulsar is now reborn as a millisecond pulsar and once again is seen to emit as a radio pulsar.[1] In what follows we do not consider the NSs which are millisecond pulsars or those seen in X-ray binaries as this standard evolutionary picture sees these two populations as originating from 'normal' radio pulsars. Their birthrates should thus be accounted for in the pulsar birthrate. However we note that if some NSs in X-ray binaries did not originate from the normal radio pulsars the problem outlined below may be emphasised even further.

RRATs

As we outlined in Sect. 2.3.2, eleven RRAT sources were discovered in 2006, showing sporadic single pulses in the radio. Underlying periodicities have been identified in all of these sources and in three cases period derivatives have been measured. We can place the three sources with known period derivative on the $P - \dot{P}$

[1] This transition from X-ray binary to millisecond pulsar has recently been observed in PSR J1023+0038 [2].

Table 3.1 Estimated birthrates in units of NSs per century for the different populations of NSs

β_{PSR}	PSR	RRAT	XDINS	Magnetar	CCO	Total	CCSN
FK06†	2.8 ± 0.5	$5.6^{+4.3}_{-3.3}$	2.1 ± 1.0	$0.3^{+1.2}_{-0.2}$	$0.5^{+0.5}_{-0.3}$	$11.3^{+7.5}_{-5.3}$	1.9 ± 1.1
L+06†	1.4 ± 0.2	$2.8^{+1.6}_{-1.6}$	2.1 ± 1.0	$0.3^{+1.2}_{-0.2}$	$0.5^{+0.5}_{-0.3}$	$7.1^{+4.5}_{-3.3}$	1.9 ± 1.1
L+06*	1.1 ± 0.2	$2.2^{+1.7}_{-1.3}$	2.1 ± 1.0	$0.3^{+1.2}_{-0.2}$	$0.5^{+0.5}_{-0.3}$	$6.2^{+4.6}_{-3.0}$	1.9 ± 1.1
V+04†	1.6 ± 0.3	$3.2^{+2.5}_{-1.9}$	2.1 ± 1.0	$0.3^{+1.2}_{-0.2}$	$0.5^{+0.5}_{-0.3}$	$7.7^{+5.5}_{-3.7}$	1.9 ± 1.1
V+04*	1.1 ± 0.2	$2.2^{+1.7}_{-1.3}$	2.1 ± 1.0	$0.3^{+1.2}_{-0.2}$	$0.5^{+0.5}_{-0.3}$	$6.2^{+4.6}_{-3.0}$	1.9 ± 1.1

The top row are the most likely values whereas the following rows give the lower limit pulsar current estimates for each of the pulsar current analyses. The Galactic electron density model that each analysis uses is denoted by either: $\dagger$ = NE2001 or $\star$ = TC93

diagram and estimate their surface magnetic field strength using Eq. 2.21. The magnetic field strengths are in the range of the normal radio pulsars ($\sim$few $\times\ 10^{12}$ G) except for J1819–1458 which lies between the normal pulsars and the magnetars with $B = 5 \times 10^{13}$ G.

XDINSs

The XDINSs are a small group of radio-quiet, close-by ($\sim$100 pc) X-ray pulsars situated in the Gould Belt, a local, partial ring of stars which includes the Sun [49]. XDINSs were originally discovered over a decade ago [62] with seven sources now known (sometimes referred to as "The Magnificent Seven"). XDINSs have thermal X-ray spectra with $kT = 50$–100 eV and show X-ray pulsations with periods in the range $\sim$3–11 s [21]. All seven sources have measured periods with three well known period derivatives and upper limits for three more (see Tables 3.1 and 3.3 in Haber [22], Tiengo and Mereghetti [54], van Kerkwijk and Kaplan [56]). However the upper limits determined are $1-2$ orders of magnitude higher than the three well known $\dot{P}$ values so may not be very constraining. We can place the three sources with known $\dot{P}$ on the $P - \dot{P}$ diagram and can infer $B \sim 10^{13}$ G in the standard way. These three sources lie just below the magnetars.

The X-ray spectra of XDINSs can be fit well with a single blackbody and interestingly do not require a power-law (e.g. synchrotron) component which suggests that XDINSs do not have magnetospheres. Also, as for RRATs, there are observed spectral features, which may be due to proton-cyclotron lines in a strong magnetic field [22]. We note that there is much current work underway, searching for RRAT-like bursty emission from XDINSs. However, no emission has been found above a flux density limit of ~10 μJy [29] from 820-MHz observations with the Green-Bank Telescope. In addition there has been no detection with GMRT at 320 MHz (B. C. Joshi, private communication). Searches are also underway using the Parkes telescope at 1.4 GHz (A. Possenti, private communication). These non-detections suggest that the identification of XDINSs as nearby RRATs [48] might be incorrect.

CCOs

Another small group of isolated NSs are the CCOs. These are isolated, non-variable point sources associated with supernova remnants (SNRs), seen in thermal X-rays without optical or radio counterparts. CCOs have low X-ray luminosities and do not have associated pulsar wind nebulae which suggests that these stars are neutron stars which are not active as pulsars. There are currently seven confirmed CCO sources[2] and about four candidate CCOs [12, 63] and deep multi-wavelength observing campaigns have been undertaken to search for more sources in nearby supernova remnants [25, 26].

Magnetars

It is thought that both Soft Gamma Repeaters (SGRs) and Anomalous X-ray Pulsars (AXPs) belong to the magnetar class of NSs [65]. Magnetars are believed to be isolated X-ray pulsars with strong magnetic fields ($10^{14}-10^{15}$ G) and periods in the range 2–12 s, and were, until recently, thought to be radio-silent. However transient radio emission has been detected from two AXPs—XTE J1810–197 [4] and 1E 1547.0–5408 [5] with both sources showing flat radio spectra. This is different to what is seen in normal radio pulsars (see Sect. 3.2). Magnetic fields strengths, inferred again from the observed spin and spin-down rates, are shown for magnetars in the same $P-\dot{P}$ diagram in Fig. 3.1.

3.3 Birthrates

The Core-Collapse Supernova Rate

Recently, the CCSN rate (for Type Ib, Ic and Type II SNe) has been determined from measurements of γ-ray radiation from ^{26}Al in the Galaxy [13]. Quantifying this γ-ray emission allowed the authors to weigh the amount of ^{26}Al in the Galaxy, as each CCSN expels a well known yield of ^{26}Al. Assuming an initial mass function (IMF), as defined in Sect. 2.1, a Scalo IMF ($d\log\Phi/d\log m = -2.7$ for high masses), they inferred the Galactic CCSN rate to be

$$\beta_{\rm CCSN} = 1.9 \pm 1.1 \text{ century}^{-1}. \tag{3.2}$$

[2] Some authors also define 1E 161348-5055, the source associated with the RCW 103 supernova remnant as a CCO. However here we follow the definition of Halpern and Gotthelf [23] and take CCOs as having steady X-ray flux. RCW 103 has variable X-ray flux.

As a consistency check, we integrate the IMF to compute,

$$\beta_{\rm CCSN} = \frac{\rm SFR}{\langle m \rangle} f_{\rm CCSN}, \tag{3.3}$$

where SFR denotes the Galactic star formation rate, $\langle m \rangle$ is the mass expectation value and $f_{\rm CCSN}$ is the fraction of stars which end their lives in a CCSN. We adopt a star formation rate of SFR $= 4\ {\rm M_\odot\ yr^{-1}}$ [13, 51]. We use the Kroupa IMF of Eq. 2.1 to determine $\langle m \rangle$ and $f_{\rm CCSN}$ which is essentially $f_{\rm NS}$ as defined in Eq. 2.3 and Table 2.1. Using the 'standard' IMF (IMF-3 from Eq. 2.1 in Sect. 2.1) we determine a CCSN rate of as high as $\approx 1.9 \pm 0.9$ century^{-1}. While this is consistent with the rate of Diehl et al. [13], considering the effects of unresolved binaries at the high-mass end (i.e. using IMF-4) we obtain a lower CCSN rate of just $\approx 0.8 \pm 0.4$ century^{-1}. In both cases here we have assumed the error to be dominated by the uncertainty of the SFR of up to $\sim$50% [51].

Radio Pulsars

The most thoroughly studied NS population are the radio pulsars. A recent estimate of their birthrate and the number in the Galaxy was performed by Lorimer et al. [33] (L+06 from herein), using 1008 non-recycled pulsars detected in 1.4-GHz surveys with the Parkes telescope (the PMPS and the Parkes High-Latitude Survey). They determined population details for sources above a 1.4-GHz radio luminosity threshold of 0.1 mJy kpc^2. This was done using a 'pulsar current' analysis [44, 59]. The basis of this idea is that the Galactic pulsar population is stationary, i.e. assuming the lifetimes of pulsars are much shorter than the age of the Galaxy we observe a pulsar population which is statistically the same at whatever time we observe during the lifetime of the Galaxy. Considering a distribution function $f(P, \dot{P})$ for $P - \dot{P}$ phase space, the number of pulsars is thus $\mathcal{N} = \int f(P, \dot{P}) dP d\dot{P}$. Regardless of our spin-down law, the distribution function must obey a continuity equation of the form:

$$\frac{\partial f}{\partial t} + \sum_{i=1}^{2} \frac{\partial}{\partial x_{\rm i}} (f v_{\rm i}) = S, \tag{3.4}$$

where the divergence term sums over both coordinates of our phase space. Here a position in phase space is $(P, \dot{P})$ so that a velocity is $(\dot{P}, \ddot{P})$, and as we have already assumed stationarity the first term vanishes. For our case we have:

$$\frac{\partial}{\partial P} (f\dot{P}) + \frac{\partial}{\partial \dot{P}} (f\ddot{P}) = S(P, \dot{P}). \tag{3.5}$$

We can see that we have a conserved current, but we can concentrate on the current in the P direction[3] by marginalising $S(P, \dot{P})$ over $\dot{P}$ to get $S(P) = \int S(P, \dot{P})d\dot{P}$ which gives

$$\frac{d}{dP}\left(\int f\dot{P}d\dot{P}\right) + \frac{d}{d\dot{P}}\left(\int f\ddot{P}d\dot{P}\right) = S(P), \tag{3.6}$$

$$\Rightarrow \frac{d}{dP}\left(J_{\mathrm{P}}(P)\right) = S(P), \tag{3.7}$$

where the derivative with respect to $\dot{P}$ vanishes as the integral is a function of P only.[4] This means that, at a period P, we have a current of $J_{\mathrm{P}}(P) = \int f\dot{P}d\dot{P}$ pulsars per unit time flowing towards higher periods. To connect this with observations we consider the distribution $\rho(P, \dot{P}, L)$ which is $f(P, \dot{P})$ per unit luminosity. Thus our conserved current is:

$$J_{\mathrm{P}}(P) = \int\int \rho(P, \dot{P}, L)\dot{P}d\dot{P}dL. \tag{3.8}$$

The true density $\rho(P, \dot{P}, L)$ is related to the observed density $\varrho(P, \dot{P}, L)$ by a scaling factor, $\xi(L)$, describing the detectability of a pulsar with luminosity L and by a beaming factor, i.e. $\rho(P, \dot{P}, L) = f_{\mathrm{beam}}^{-1}\xi(L)\varrho(P, \dot{P}, L)$. In practise the observed distribution is discrete, i.e. for each of N_{PSR} pulsars we know values for P, $\dot{P}$ and L so that

$$\rho(P, \dot{P}, L) \approx \sum_{i=1}^{N_{\mathrm{PSR}}} f_{\mathrm{beam}}^{-1}\xi(L)\delta(P - P_i)\delta(\dot{P} - \dot{P}_i)\delta(L - L_i). \tag{3.9}$$

Averaging $J_{\mathrm{P}}(P)$ within bins of with ΔP we calculate $J_{\mathrm{P,ave}}(P) = (\Delta P)^{-1} \int J_{\mathrm{P}}(P)dP$. Substituting Eq. 3.9 into this and integrating out the delta functions we get that the current in a bin defined between P_1 and P_2 is

$$J_{\mathrm{P,ave}}(P) = \frac{1}{\Delta P}\sum_{i=1}^{N_{\mathrm{PSR,bin}}} f_{\mathrm{beam}}^{-1}\xi(L)\dot{P}_i, \tag{3.10}$$

where $\Delta P = P_2 - P_1$ and P is the centre of the bin. If there were no selection effects then the current in a period bin would simply be the sum of the period derivatives of pulsars within that bin. We can see that $J_{\mathrm{P}}(P)$ equals the birthrate minus the deathrate in the period range $0 - P$. In particular, if pulsars are all born at some $P < P_1$ and all die at some $P > P_2$ the plateau value of $J_{\mathrm{P}}(P)$ between P_1 and P_2 equals the pulsar birthrate. However, even assuming such a plateau and having corrected for the selection effects as much as possible, the analysis will yield a lower limit to the pulsar birthrate, as the observed pulsar sample is flux limited.

---

[3] We could also separate out the current in the $\dot{P}$ direction but this would require a knowledge of $\ddot{P}$ which we do not have.

[4] And we have used the Leibniz Integral Rule, i.e. $\frac{d}{dx}\int f(x, y)dy = \int \frac{\partial}{\partial x} f(x, y)dy$.

The results obtained have model dependencies, both on the Galactic electron density model and on the pulsar beaming fraction model. The current best model for the electron density is the NE2001 model [9] and L+06 adopted this as well as the Tauris and Manchester [52] beaming model for their calculations. L+06 determine a birthrate of $\beta_{PSR} = 1.38 \pm 0.21$ century^{-1} and $N_{PSR} = 155000 \pm 6000$. This result is consistent with the earlier work of Vranesevic et al. [61] (V+04 from herein) which used 815 non-recycled PMPS pulsars to determine $\beta_{PSR} = 1.58 \pm 0.33$ century^{-1} and $N_{PSR} = 106600 \pm 11700$, in this case determined above a higher threshold of 1 mJy kpc^2. In both cases, as is common to allow for direct comparisons with older results, the now superseded TC93 [53] electron density model was also used and for both analyses this produced lower birthrate estimates.

More recent work by Faucher-Giguère and Kaspi [15] (FK06 from herein) yields a much higher pulsar birthrate of $\beta_{PSR} = 2.8 \pm 0.5$ century^{-1}. The approach of this analysis is different—the authors model the birth properties of pulsars (velocity distributions, magnetic fields and detectability in the PMPS and Swinburne Multibeam surveys) from the observational data before performing Monte Carlo simulations to evolve the initial population to obtain the observed pulsar sample. The quoted birthrate is the average of 50 runs of their simulations. While it is twice as large as that provided by the pulsar current analyses the results are entirely consistent as the pulsar current analysis is, as we outlined above, to be interpreted as providing a reliable lower limit to the pulsar birthrate. This FK06 value is currently the best pulsar birthrate estimate available.

RRATs

The estimated number of RRATs is $N_{RRAT} \gtrsim 2-4 \times 10^5$ [38] and therefore even higher than that of the radio pulsar population (see Sect. 3.5 for discussion of the various parameters on which this estimate depends). However, the determination of N_{RRAT} is obviously based on a very small sample of sources. In order to account for this uncertainty, we will use the following parameterisation in our computations, i.e. $N_{RRAT} = \gamma N_{PSR}$ where we take $\gamma \sim 1-3$.

It is important to realise the following caveat when interpreting this estimate for the total number of RRATs. The fact that it appears to be larger than that of pulsars does not necessarily imply that a NS is more likely to be a RRAT than a pulsar. This would assume that the physical mechanisms for the emission of the RRAT radio bursts is identical to that of regular pulsars. There is no reason to assume this, especially as RRAT spectra and polarisation properties are as yet unknown. Emission criteria (which may represent certain 'active' areas on the $P - \dot{P}$ diagram) for RRAT and pulsar emission may be different and the respective 'death-lines' may also be different, so that the duration of RRAT and pulsar emitting phases would not be the same either. Here lies the advantage of considering birthrates (e.g. pulsar current analyses) rather than absolute population numbers [48].

However, as we do not have a reliable age estimator for RRATs, we are forced to assume similar active lifetimes for RRATs and pulsars to work out birthrates from population estimates. We could conceivably use temperature as a measure of age (see Sect. 3.5) but as there is just one RRAT with known temperature we follow Popov et al. [26] who have argued that if RRATs are rotating NSs with pulsar-scale magnetic fields then the active lifetime of pulsars $\tau_{PSR} \approx N_{PSR}/\beta_{PSR} \sim 5 \times 10^6$ yr would be similar to that of RRATs, τ_{RRAT}. This holds provided the initial spin periods of RRATs and pulsars are within a factor of a few of each other. With the conclusion of approximately equal time-scales and $N_{RRAT} = \gamma N_{PSR}$, this implies a RRAT birthrate of $\beta_{RRAT} \approx \gamma\beta_{PSR}$ [26]. Thus if we take $\gamma \sim 2$ we have an indicative RRAT birthrate of $\beta_{RRAT} \sim 2.8 \pm 1$ century^{-1} considering the pulsar current analyses, or as large as $\beta_{RRAT} \sim 5.6 \pm 1$ century^{-1} considering the FK06 result.

XDINSs

The birthrate for XDINSs has recently been estimated by Gill and Heyl [19]. These authors performed a population synthesis for XDINSs based on the seven sources detected in the ROSAT All-Sky Survey [60]. A limiting volume for OB progenitor stars was determined and then compared to the actual number of OB stars detected in this volume in the survey, to determine the relevant scalings. The authors then use an age estimate to find birthrates from the simulated number of sources and determine $\beta_{XDINS} \sim 2.1 \pm 1$ century^{-1}. The age estimate is arrived at simply from averaging the characteristic age for the two XDINSs which then had well-known $\dot{P}$'s($\approx$1.5 and $\approx$1.9 Myr), and earlier estimates for their NS cooling ages ($\sim$0.5 Myr). The result is consistent with a recent lower estimate of $\beta_{XDINS} \sim 1$ century^{-1} made by Popov et al. [48] which used a NS cooling age of $\tau_{XDINS} \approx 1$ Myr.

CCOs

An estimate of the CCO birthrate has been made by Gaensler et al. [18], the question not having been revisited since. They considered the six nearby (within 3.5 kpc) CCOs which have ages less than 20 kyr. Here the ages are estimated from the supernova remnant expansion times. Extrapolating this to the entire Galaxy gives a birthrate of $\beta_{CCO} \approx 0.5$ century^{-1}. Below, we include this contribution, adding an ad hoc uncertainty factor of two, but noting that excluding CCOs does not change any of the conclusions.

Magnetars

Magnetar birthrates have typically been determined using two different methods of age estimation, required to convert simulated populations to birthrates. The first method uses spin-down age estimates for magnetars, as done by Kouveliotou et al.

[30] to determine an SGR birthrate of $\beta_{\mathrm{SGR}} \approx 0.1$ century^{-1} which we consider as a lower limit for the magnetar birthrate [65]. Similarly, an AXP birthrate was calculated by Gill and Heyl [19] using the same population synthesis methods used for the XDINSs, obtaining $\beta_{\mathrm{AXP}} \sim 0.2 \pm 0.2$ century^{-1}. Another recent determination of $\beta_{\mathrm{magnetar}} = 0.15-0.3$ century^{-1} has been reported by Ferrario and Wickramasinghe [17]. The second means by which magnetar age estimates can be obtained, involves using ages of supernova remnant associations of SGRs and AXPs. These have yielded slightly smaller estimates [58] as the supernova remnant ages tend to be longer than the spin-down ages resulting in a smaller birthrate.

Due to these small magnetar birthrate estimates relative to the other populations of NSs, one might think that we can safely neglect the magnetar contribution to Eq. 3.1. However, we note the possibility that if, for example, magnetars experience magnetic field decay (as considered by Arras et al. [3] and by Colpi et al. [7]) the true age is smaller than the characteristic age. This seems an important consideration as magnetar emission is thought to be powered by decaying magnetic fields [65]. This would imply a higher birthrate, possibly as high as $\sim$2 century^{-1} for AXPs [19]. In addition, larger magnetar birthrate estimates have been reported recently by Muno et al. [39]. These authors studied 947 archival observations from *XMM Newton* and *Chandra*. From the 7 magnetars detected they determine the most likely number of Galactic magnetars considering the small fraction of the sky covered in these observations. They obtain, separately, birthrates for persistent AXPs, transient AXPs as well as a small contribution from SGRs yielding a large magnetar birthrate of $\beta_{\mathrm{magnetar}} = 2.6^{+5.0}_{-1.5}$ century^{-1}. This however assumes lifetimes of 10^4 yr (see Fig. 3.1) for each of these sub-populations. As the lifetime for transient AXPs is very uncertain it is possible that their lifetime is larger by an order of magnitude. In this case the persistent AXPs give the most reliable magnetar birthrate of $\beta_{\mathrm{magnetar}} = 0.6^{+0.9}_{-0.3}$ century^{-1}.

It is not clear if the question of beaming has been considered in the estimates of Muno et al. [39] but we should not necessarily expect magnetar emission to be isotropic. In this case, only magentars beamed toward us will have been detected. However, judging from the observed pulse shapes (e.g. Woods and Thompson [65]), we assume that the beaming fraction is larger than for radio pulsars, so that the effect of beaming may not be quite as significant as for pulsars. Nevertheless noting that the derived values may represent a lower limit—we proceed by neglecting this extra beaming factor—and, considering all the estimates reviewed here we adopt a conservative magnetar birthrate of $\beta_{\mathrm{magnetar}} \approx 0.3^{+1.2}_{-0.2}$ century^{-1} where our extended error bars allow for the potentially much higher values suggested by the Muno et al. [39] study.

3.4 Too Many Neutron Stars?

The birthrates for each of the NS populations are summarised below in Table 3.1 and Fig. 3.2. It appears that the CCSN rate cannot sustain all the separate NS populations. In a previous consideration of this question by Popov et al. [48], XDINSs and RRATs

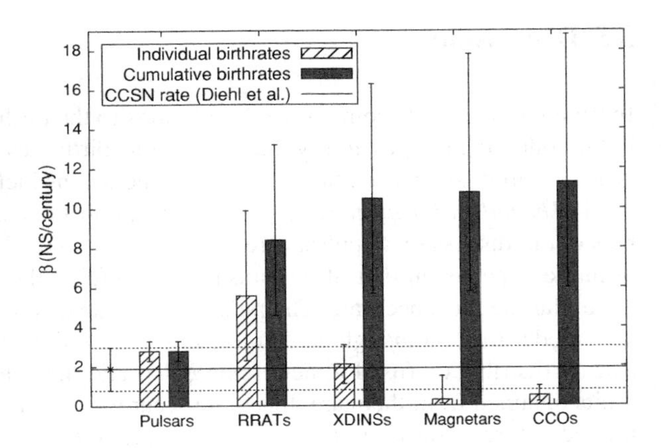

Fig. 3.2 Estimated birthrates for the individual NS populations (*hatched*), the cumulative number (*grey*) with associated errors. The CCSN rate is shown as a horizontal line (*solid*), as are its error bars (*dashed*)

were identified as a single NS population, so that only one birthrate contribution was taken, i.e. that of the XDINSs. Moreover, the magnetar contribution was neglected and an XDINS birthrate was assumed such that $\beta_{\mathrm{XDINS}} \approx \beta_{\mathrm{PSR}}$. In addition, Popov et al. [48] used the lower limit pulsar birthrate of V+04 to be the pulsar birthrate, a result since superseded by the work of FK06. In this picture, where XDINSs are identified with nearby RRATs, the total birthrate is $\beta_{\mathrm{total}} = 2 - 4$ century^{-1} which is barely consistent with the CCSN rate. However, including the magnetars, allowing for separate RRAT and XDINS contributions and using the more accurate pulsar birthrate of FK06, Eq. 3.1 cannot be satisfied with the estimates from Table 3.1. This seems to be the case even if we assume the highest CCSN rate allowable within the uncertainties (i.e. $\beta_{\mathrm{CCSN}} = 3$ century^{-1}) while at the same time allowing for the lowest total required NS birthrate, $\beta_{\mathrm{total}} = 6.2$ century^{-1}. It seems that the number of NSs produced via CCSNe is not sufficient.

We can just about reconcile the rates if we choose the highest allowable CCSN rate and the lowest allowable total NS birthrate from the L + 06 result using the TC93 electron density model (see Table 3.1). However, as we discussed earlier, the pulsar current results are lower limits and the NE2001 model is often considered to be a more accurate model than TC93 [9, 10]. From looking at Fig. 3.2 we are left to conclude that either the individual NS birthrates are over-estimated (or the uncertainties in these values are under-estimated) or the assumption of distinct NS populations needs to be revised. To reconcile the values within the errors would require the RRAT and XDINSs errors (which recall are the most uncertain) to each be under-estimated by a factor of 2. If this is not the case then it would seem that Eq. 3.1 is not satisfied. Taking this at face value implies that there are too many NSs in the Galaxy. We will discuss the nature of this potential NS 'birthrate problem' in the following section.

3.5 Discussion

In trying to determine some possible solutions to the birthrate problem we consider in the following the possibility that the various birthrates are incorrect or that there is an evolutionary answer. Some possible conclusions include:

(1) *The pulsar birthrate is wrong*. The pulsar birthrate is the most crucial component of our discussion as pulsars are the most well-studied population and the RRAT birthrate depends on that of the pulsars. Thankfully, the pulsar birthrate estimates are by far the most accurate. The pulsar current analyses make no assumptions and are "model free", although are subject to the uncertainties in the Galactic electron density distribution (used to determine $\xi(L)$) and the beaming fraction. The lower limits obtained from them are thus quite secure. In order to compensate for the flux limited nature of these studies, we would need to choose a functional form for the luminosity (depending on P and $\dot{P}$) but the inclusion of such a correction *can only increase the determined birthrate*.

The work of FK06 models this luminosity evolution across the $P - \dot{P}$ diagram as well as many other birth properties (modelled from the observed pulsar population). The analysis did assume magnetic dipole spin-down of pulsars but allowed for magnetic field decay as well as drawing braking indices from a uniform distribution in the range $n \in [1.4, 3.0]$ (note that the few measured braking indices are found to lie in the range 1.4–2.9, see [32, 35] and references therein).

Another uncertainty for pulsars is the beaming fraction. An indication of this may be the recent discovery of a pulsar with an extremely small duty cycle [28]. This pulsar has a beaming fraction of just 0.04% or only 0.14° of longitude. Usually, we would expect the minimum pulse width (for an orthogonal rotator) for this pulsar period of $P = 91$ ms to be given by $W_{\rm min}(h) \sim 8.2°(h/10\ {\rm km})^{1/2}$ [34] where h is the emission height and we have assumed β, the impact parameter, to be small. What is observed is a pulsar that is narrower by a factor of $\sim$10. It is possible that the pulse represents a cut at the very edge of the conical beam but this seems to be at odds with the two observed distinct components in the pulse profile [28]. Pulsars with pulse widths this narrow therefore raise the question of whether or not our beaming fraction estimates are accurate. If they are in fact over-estimates then there may be many more pulsars which we do not see.

In summary, taking the pulsar current analysis to provide a reliable lower limit it seems indeed reasonable to take a pulsar birthrate of $\beta_{\rm PSR} = 2$ century^{-1} as being quite conservative when the many low-luminosity pulsars are included.

(2) *The RRAT birthrate is wrong and hugely over-estimated*. The RRAT birthrate depends on the RRAT population estimate being correct. This is based on assumptions that the Galactic distribution of RRATs follows that of pulsars, on assumptions about the impact of man-made radio frequency interference (RFI) during the PMPS observations, beaming and Galactic electron distribution models used and on RRAT burst rate estimates. The full expression is given in Eq. 2.37. All of these effects are treated conservatively but to really improve the accuracy of the estimate, we need to discover many more sources. This will enable us

to accurately determine the factors in this equation. The 'on'-factor can be best constrained with accurate RRAT burst rates for a larger population of RRATs. The RFI factor is more difficult to quantify but recently we have made progress in this regard with the development of a new RFI removal scheme which is capable of removing the vast majority of terrestrial RFI. The RFI removal means that we will be able to essentially ignore the $f_{\rm RFI}$ factor (i.e. $f_{\rm RFI} \rightarrow 0$). These developments have motivated a complete re-processing of the PMPS data which is fully described in Chap. 4.

An uncertainty in the beaming fraction of radio pulsars also affects RRATs, as we have, as the best available assumption, adopted the beaming model of pulsars. This is empirically determined from measurements of low period ($P < 2$ s) pulsars so that we are extrapolating to long periods when applying the model to RRATs. If the beams are narrower, then this increases the estimate for the number of RRATs. The RRAT lifetime is another uncertain parameter. If one were to propose a much larger active lifetime than that proposed by Popov et al. [48] this would reduce the implied RRAT birthrate. However even if this were an order of magnitude larger (i.e. $\sim$50 Myr) the birthrate problem would remain, although less emphatic. Conversely we note that the RRATs seem to have higher $\dot{P}$ values than most pulsars and thus may evolve to higher periods (i.e. towards the death valley) more quickly than pulsars which would then imply a shorter lifetime than that assumed above. Noting all of these caveats, it does not seem unreasonable to take, as a conservative estimate, $\beta_{\rm RRAT} = 2 - 6$ century^{-1} for $\gamma = 1 - 3$ as before.

(3) *The XDINS birthrate is wrong and hugely over-estimated.* As in the case of RRATs, this could be due to the small sample size of this population. Furthermore, some of the XDINS birthrate estimates assume spin-down ages to be an accurate age estimate and there are only three XDINSs with well known $\dot{P}$. Additionally, the estimate of Gill and Heyl [19] also depends on the used Galactic $N_{\rm HI}$ model, i.e. the authors applied an exponential model (ignoring any warp or spiral arm components). Long-term monitoring of the known XDINSs can yield exact period derivatives which in turn will give us accurate characteristic ages which, along with NS cooling ages, may enable us to determine the true age of XDINSs. Certainly, however, as for RRATs, the best way to improve the estimates is to increase the known population of XDINSs. Recent work to determine where best to search for these elusive sources, has pointed towards the Cygnus-Cepheus region behind the Gould Belt [50].

(4) *A possible evolution from different types of NSs into others.* We consider here the possibility that pulsars, RRATs and XDINSs might be different evolutionary stages of a single class of object. If this is the case, we need only take one birthrate for these populations into balancing the CCSN rate, i.e. the birthrate of the earliest stage in this cycle. To determine the direction of this evolution requires a reliable age estimator. As all of the considered NSs are isolated objects, we expect that they simply slow down as they age, so that the longer periods of XDINSs (3–11 s), compared with those of the RRATs (0.7–7 s), imply RRATs to be younger than XDINSs. Similarly, the even lower periods of isolated pulsars (0.03–8.5 s) might imply that the evolutionary track in question is pulsar

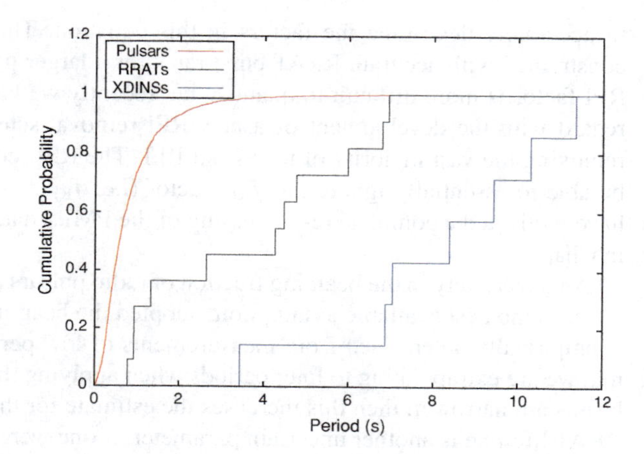

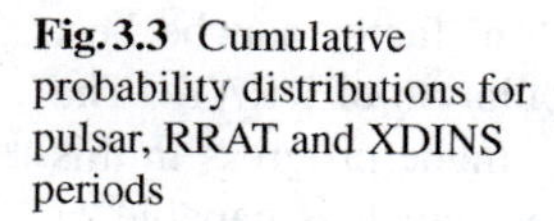
Fig. 3.3 Cumulative probability distributions for pulsar, RRAT and XDINS periods

→ RRAT → XDINS. To investigate this possibility, we performed Kolmogorov–Smirnov (K–S) tests comparing the cumulative period probability distributions (see Fig. 3.3). The probability that the pulsar and RRAT distributions are drawn from the same parent distribution is found to be 0.02%. However, we believe the comparison to be

unfair due to the large difference in distribution sizes (1500 and 11). To test this we randomly selected 20 pulsar periods from their period distribution and compared these with the RRAT distribution over many iterations. The resulting probabilities vary largely with ∼17% of iterations showing probabilities below 1% but ∼18% of iterations showing probabilities above 20%. Next we compared the median periods of the randomly selected pulsar samples to that of the RRATs. This value is stable with the median pulsar period being 615 ms over 10,000 iterations. The RRAT median period is $\sim 7\sigma$ above this value considering 16 known RRATs (the 11 original sources plus 5 newly detected sources, M. McLaughlin, private communication) and $\sim 20\sigma$ considering only the published sources. From this we conclude that RRAT periods are intrinsically longer than those of the pulsars.

Comparing the RRAT and XDINS distributions using the K–S test gives a probability that these two distributions are drawn from the same parent distribution of only 2% but not low enough to reject this possibility given the small numbers in each category ($N_{\mathrm{RRAT}} = 11$, $N_{\mathrm{XDINS}} = 7$). If we nevertheless assume these distributions to be different, we can estimate the time needed for the RRAT period distribution to evolve to the XDINS period distribution by comparing the average period and the period derivatives. The average periods are, respectively, $\langle P_{\mathrm{RRAT}} \rangle = 3.6$ s and $\langle P_{\mathrm{XDINS}} \rangle = 8.1$ s. Using the average RRAT period derivative of $\langle \dot{P} \rangle = 2 \times 10^{-13}$ and assuming that it is constant with time, we estimate an evolutionary time of

$$t = \frac{\langle P_{\mathrm{XDINS}} \rangle - \langle P_{\mathrm{RRAT}} \rangle}{\langle \dot{P}_{\mathrm{RRAT}} \rangle} \sim 0.7 \text{ Myr.} \tag{3.11}$$

Another indicator of age is temperature, assuming that no significant heating occurs during the life of any of these sources. Assuming NSs are born with the same temperature and then cool along a NS cooling curve, we can determine the age if the surface temperature can be measured.[5] However, NS cooling curves are not very well constrained (as they often depend on the unknown NS EoS, e.g. Yakovlev and Pethick [66]; as discussed in Chap. 2 this is difficult to determine) so that the exact age-temperature relationship is not known. Still, we can safely expect an older star to be cooler than a young NS, so that we might suppose XDINSs to be cooler than RRATs as XDINSs have $T \approx 0.7 \times 10^6$ K [66] and RRAT J1819−1458 has $T \approx 1.6 \times 10^6$ K, lending tentative support to our conclusion based upon periods.

If some/all RRATs do evolve into XDINSs it means the XDINS birthrate can be removed from consideration in balancing Eq. 3.1. We note that a NS evolution from RRAT to XDINS suggests a path on the $P - \dot{P}$ diagram (along, say, an $n = 3$ line) which starts in the region of the high-B radio pulsars. Such an evolution where RRATs and XDINSs are evolutionary states reached by some of the normal radio pulsars means both of these birthrate contributions can be removed from Eq. 3.1 and the birthrate problem is much less severe and is solved within the errors of $\beta_{\rm PSR}$ and $\beta_{\rm CCSN}$.

This of course does not include the magnetar contribution. If the magnetar birthrate is in fact small, it is not necessary to include magnetars in an evolutionary scenario to solve the problem but nonetheless we consider it. Comparing the magnetar period distribution to that of RRATs and XDINSs suggests that the XDINSs and magnetar distributions are quite similar. A K–S test finds their distributions to be the same with a 92% probability. This might suggest that the XDINS and magnetar evolutionary end states come from the same initial population. Such a scenario with some high-B radio pulsars evolving towards the magnetar region of the $P - \dot{P}$ diagram involves increasing B. While at first this may seem counter-intuitive we note that Eq. 2.21 describes the surface magnetic field which could indeed increase with time. In fact, this is suggested by long-term observations of the Crab pulsar and PSR B1737−30 [31, 35], and by the measured braking indices from young pulsars which have $n < 3$.

(5) *An unknown NS formation process*. Another (even more drastic) solution might be that there is some other unknown mechanism of forming NSs besides in a CCSN. This might well be required even if there is an evolutionary answer to the birthrate problem because, as we have noted, the number of pulsars alone is pushing the allowed limit. The only other known mechanisms for forming NSs are electron-capture SNe (ECSN) [40, 41] and an accretion induced collapse (AIC) [20].

Stars with initial masses[6] in the range 8–11 $\pm$ 1 $M_{\odot}$ form stars with He cores of mass 1.6–3.5 $M_{\odot}$ [55]. Burning in the core produces an O–Ne–Mg core which,

[5] We note the extra downward emission above the surfaces of pulsars will act as a heating mechanism of the polar cap. This scenario does not apply for RRATs and XDINSs if they are 'off' most and all of the time.

[6] The exact range depends on how convection is treated, e.g. large convective overshooting and low metallicity make the limit go from 12 to 10 $M_{\odot}$ [14], hence the uncertainty in the high mass value.

as the temperature does not become sufficient to ignite Ne, becomes degenerate.[7] If the O–Ne–Mg core can grow, from He shell burning, it will eventually reach a critical density of $\approx 4.5 \times 10^{12}$ kg m^{-3} [45] where electron capture onto Mg (and hence Ne) can occur. This removes the electrons and hence the degeneracy pressure supporting the star and results in a core-collapse to a NS—this is an ECSN. In binary systems, the hydrogen envelope of the star is lost quickly (long before the AGB phase) and the He core can progress towards an ECSN. In fact the lighter companion, pulsar B, in the double pulsar binary seems to have been formed in this way. The critical density corresponds to a (baryonic) mass in the range 1.366–1.375 $M_{\odot}$, which, as it turns out, equals the (gravitational) mass of pulsar B plus its binding energy, to high precision [16].

The birthrate problem involves isolated stars, which tend to lose their He shells in the second dredge up, and lose their H envelopes due to heavy wind losses in the AGB phase. This means that the O–Ne–Mg core cannot grow to the critical value for ECSN and an O–Ne–Mg white dwarf (WD) remains. This picture says that ECSN are only possible in binaries [46, 55] but this may not be strictly true. Recent work has shown that there is a narrow mass range of $9.00 - 9.25\ M_{\odot}$ [47] where ECSN can occur in isolated stars. Invoking IMF-4 from Eq. 2.1 this gives the fraction of NSs formed in ECSN to be:

$$\frac{\text{ECSN}}{\text{ECSN} + \text{CCSN}} = \frac{\int_{9.00}^{9.25} m^{-2.7} dm}{\int_{9.00}^{25.00} m^{-2.7} dm} \approx 5.5\%. \tag{3.12}$$

So ECSN increase the number of Galactic NSs but only by a small amount. This may even be an overestimate, as there are no pulsars yet observed with low space velocities [24], symptomatic of ECSN, as observed in binary pulsars formed in this way [55].

AIC is a process naturally only occurring in binary systems. Here accretion onto a C–O WD can, if sufficiently slow, make it collapse to a O–Ne–Mg WD. If the slow accretion persists we get an ECSN and a NS is produced. The accretion rate must be slow so that the star is not completely unbound in a Type Ia supernova, and so that it does not lose the accreted mass in a series of nova explosions. We note that this scenario is different to the binary ECSN described earlier as the mass of the O–Ne–Mg core in that case grew through burning of its He shell but here the WD can be initially less massive (i.e. a C–O WD) and grow towards ECSN. Clearly, by definition AIC does not apply to isolated stars, and does not eject stars from binary systems (or globular clusters, see e.g. Pfahl et al. [43]) so cannot resolve the birthrate problem. It would seem that another, as yet unknown, NS formation mechanism would be required.

[7] In more massive stars the core does not become degenerate, Ne burns, and eventually an Fe core forms and we get a CCSN.

3.6 Conclusions

In this work we have presented a detailed and critical review of the current state of birthrate calculations for a number of NS manifestations. Based on this review, we suggest that the current CCSN rate cannot explain the birthrates of the various NS populations. Unless new birthrate estimates emerge which differ vastly from those discussed here, we have a birthrate problem. If this is the case we favour an evolutionary interpretation where radio pulsars, RRATs and XDINSs (and possibly also magnetars) are different evolutionary stages of the same object. Another possible, more exotic solution would be the existence of some, as yet unknown, mechanism for forming NSs. While the determined birthrates are uncertain for XDINSs, RRATs and magnetars we consider the currently claimed uncertainties, if in fact they are correct, not to be large enough to convincingly remove the described problem.

To truly advance in answering these questions, it is essential to find a significantly larger number of sources to increase the known populations of RRATs, XDINSs and magnetars. Searches to this end are underway and many more are planned (in particular see Chap. 4). There are prospects for discovering XDINSs and magnetars from future high energy observatories. These include the International X-Ray Observatory (which has now superseded the planned XEUS and Con-X missions) as well as the Fermi gamma-ray space observatory. Radio surveys include the ongoing P-ALFA survey at Arecibo [11], the planned transients and pulsar observations with LOFAR, and the new Parkes and Effelsberg multi-beam all-sky surveys. Ultimately, a Galactic census of pulsars with the Square Kilometre Array (SKA) should improve our knowledge of Galactic NSs phenomenally with the expected detection of $\sim$20000 new pulsars [8]. Together, the SKA and LOFAR will monitor the Northern and Southern hemisphere essentially continuously. Even though the spectrum of RRATs and hence their discovery potential at low frequencies still has to be assessed, both telescopes should allow us to potentially observe most RRATs in the Galaxy that are beaming towards Earth. In other words, SKA observations will, with certainty, establish the relative population numbers for RRATs and pulsars, confirming or rejecting the results of this study. Extremely valuable information will, however, be already available much sooner with the help of LOFAR [57].

Final Thoughts

Here we have described the NS birthrate problem as it stood in 2008[8] and put forward a number of possible solutions. An evolutionary link between the populations could solve this problem. Such a solution highlights the fact that the evolution of neutron stars post-supernova is not understood and no satisfactory evolutionary picture exists. After examining the implications it seems that there is no reason for considering RRATs, and indeed the other classes of NS described here, as distinct. These

[8] In Appendix C we mention some updated supplementary information pertaining to this chapter.

investigations can be furthered by improved population estimates—something which is most feasible for RRATs. With this in mind we proceed to outline a re-processing of the PMPS, presenting the search methodology and resultant discoveries before discussing follow-up studies.

References

1. M.A. Alpar, A.F. Cheng, M.A. Ruderman, J. Shaham, Nature **300**, 728 (1982)
2. A.M. Archibald et al., Science **324**, 1411 (2009)
3. P. Arras, A. Cumming, C. Thompson, ApJ **608**, L49 (2004)
4. F. Camilo, S.M. Ransom, J.P. Halpern, J. Reynolds, D.J. Helfand, N. Zimmerman, J. Sarkissian, Nature **442**, 892 (2006)
5. F. Camilo, J. Reynolds, S. Johnston, J.P. Halpern, S.M. Ransom, ApJ **681**, 681 (2008) (astro-ph/0802.0494)
6. K. Chen, M. Ruderman, ApJ **402**, 264 (1993)
7. M. Colpi, U. Geppert, D. Page, ApJ **529**, L29 (2000)
8. J.M. Cordes, M. Kramer, T.J.W. Lazio, B.W. Stappers, D.C. Backer, S. Johnston, New Astr. **48**, 1413 (2004)
9. J.M. Cordes, T.J.W. Lazio, preprint (arXiv:astro-ph/0207156)
10. J.M. Cordes, T.J. Lazio, preprint (arXiv:astro-ph/0301598)
11. J.M. Cordes et al., ApJ **637**, 446 (2006)
12. A. de Luca, in *American Institute of Physics Conference Series*, Vol. 983, ed. by C. Bassa, Z. Wang, A. Cumming, V.M. Kaspi (40 Years of Pulsars: Millisecond Pulsars, Magnetars and More, 2008), p. 311
13. R. Diehl et al., Nature **439**, 45 (2006)
14. J.J. Eldridge, C.A. Tout, MNRAS **353**, 87 (2004)
15. C.-A. Faucher–Giguère, V.M. Kaspi, ApJ **643**, 332 (2006)
16. R.D. Ferdman et al., in *American Institute of Physics Conference Series*, Vol. 983, ed. by C. Bassa, Z. Wang, A. Cumming, V. M. Kaspi, 40 Years of Pulsars: Millisecond Pulsars, Magnetars and More (2008), p. 474 (astro-ph/0711.4927), http://adsabs.harvard.edu/abs/2008AIPC..983..474F
17. L. Ferrario, D. Wickramasinghe, MNRAS **389**, L66 (2008)
18. B.M. Gaensler, D.C.-J. Bock, B.W. Stappers, ApJ **537**, L35 (2000)
19. R. Gill, J. Heyl, MNRAS **382**, 52 (2007)
20. J.E. Grindlay, in *The Origin and Evolution of Neutron Stars*, IAU Symposium No. 125, ed. by D.J. Helfand, J. Huang (Reidel, Dordrecht, 1987), p. 173
21. F. Haberl, Adv. Space Res. **33**, 638 (2004)
22. F. Haber, Astrophysics and Space Science **308**, 171 (2007) (astro-ph/0609066)
23. J.P. Halpern, E.V. Gotthelf, ApJ **709**, 436 (2010)
24. G. Hobbs, D.R. Lorimer, A.G. Lyne, M. Kramer, MNRAS **360**, 974 (2005)
25. D.L. Kaplan, D.A. Frail, B.M. Gaensler, E.V. Gotthelf, S.R. Kulkarni, P.O. Slane, A. Nechita, ApJS **153**, 269 (2004)
26. D.L. Kaplan, B.M. Gaensler, S.R. Kulkarni, P.O. Slane, ApJS **163**, 344 (2006)
27. E.F. Keane, M. Kramer, MNRAS **391**, 2009 (2008)
28. M.J. Keith, S. Johnston, M. Kramer, P. Weltevrede, K.P. Watters, B.W. Stappers, MNRAS **389**, 1881 (2008)
29. V.I. Kondratiev, M. Burgay, A. Possenti, M.A. McLaughlin, D.R. Lorimer, R. Turolla, S. Popov, S. Zane, in *American Institute of Physics Conference Series*. Vol. 983, ed. by C. Bassa, Z. Wang, A. Cumming, V.M. Kaspi, 40 Years of Pulsars: Millisecond Pulsars, Magnetars and More (2008), p. 348, http//adsabs.harvard.edu/abs/2008AIPC..983..348K

30. C. Kouveliotou et al., Nature **393**, 235 (1998)
31. J.M. Lattimer, B.F. Schutz, ApJ **629**, 979 (2005)
32. M.A. Livingstone, V.M. Kaspi, F.P. Gavriil, R.N. Manchester, E.V.G. Gotthelf, L. Kuiper, Ap&SS **308**, 317L (2007)
33. D.R. Lorimer et al., MNRAS **372**, 777 (2006)
34. D.R. Lorimer, M. Kramer, *Handbook of Pulsar Astronomy* (Cambridge University Press, 2005)
35. A.G. Lyne, in *Young Neutron Stars and Their Environments*, vol. 1, Gaensler IAU Symposium 218, ed. by F. Camilo, B.M. Gaensler (Astronomical Society of the Pacific, San Francisco, 2004), p. 257
36. V.M. Malofeev, J.A. Gil, A. Jessner, I.F. Malov, J.H. Seiradakis, W. Sieber, R. Wielebinski, A&A **285**, 201 (1994)
37. O. Maron, J. Kijak, M. Kramer, R. Wielebinski, A&AS **147**, 195 (2000)
38. M.A. McLaughlin et al., Nature **439**, 817 (2006)
39. M.P. Muno, B.M. Gaensler, A. Nechita, J.M. Miller, P.O. Slane, ApJ **680**, 639 (2008)
40. K. Nomoto, ApJ **277**, 791 (1984)
41. K. Nomoto, ApJ **322**, 206 (1987)
42. G.G. Pavlov, D. Sanwal, G.P. Garmire, V.E. Zavlin, in *Neutron Stars in Supernova Remnants,* ed. by P.O. Slane, B.M. Gaensler. Astronomical Society of the Pacific Conference Series, vol. 271, (2002), p. 247
43. E. Pfahl, S. Rappaport, P. Podsiadlowski, ApJ **573**, 283 (2002)
44. E.S. Phinney, R.D. Blandford, MNRAS **194**, 137 (1981)
45. P. Podsiadlowski, J.D.M. Dewi, P. Lesaffre, J.C. Miller, W.G. Newton, J.R. Stone, MNRAS **361**, 1243 (2005)
46. P. Podsiadlowski, N. Langer, A.J.T. Poelarends, S. Rappaport, A. Heger, E.D. Pfahl, ApJ **612**, 1044 (2004)
47. A.J.T. Poelarends, F. Herwig, N. Langer, A. Heger, ApJ **675**, 614 (2008)
48. S.B. Popov, R. Turolla, A. Possenti, MNRAS **369**, L23 (2006)
49. W. Poppel, Fund. Cos. Phys. **18**, 1 (1997)
50. B. Posselt, S.B. Popov, F. Haberl, J. Truemper, R. Turolla, R. Neuhaeuser, A&A **482**, 617 (2008) (astro-ph/0801.4567)
51. W.W. Stahler, F. Palla, The Formation of Stars (Weinheim, Wiley-VCH, 2004)
52. T.M. Tauris, R.N. Manchester, MNRAS **298**, 625 (1998)
53. J.H. Taylor, J.M. Cordes, ApJ **411**, 674 (1993)
54. A. Tiengo, S. Mereghetti, ApJ **657**, L101 (2007)
55. E.P.J. van den Heuvel, in *The Multicolored Landscape of Compact Objects and Their Explosive Origins*, ed. by di T. Salvo, G.L. Israel, L. Piersant, L. Burderi, G. Matt, A. Tornambe, M.T. Menna, American Institute of Physics Conference Series, Vol. 924, (2007), p. 598
56. van Kerkwijk, Kaplan, APJ **673**, L163 (2008), http://adsabs.harvard.edu/abs/2008APJ...673L.163V
57. J. van Leeuwen, B. Stappers, in *American Institute of Physics Conference Series*, Vol. 983, ed. by C. Bassa, Z. Wang, A. Cumming, V.M. Kaspi, 40 Years of Pulsars: Millisecond Pulsars, Magnetars and More (2008), p. 598
58. J. van Paradijs, R.E. Taam, van den E.P.J. Heuvel, A&A **299**, L41 (1995)
59. M. Vivekanand, R. Narayan, J. Astrophys. Astr. **2**, 315 (1981)
60. W. Voges et al., A&A **349**, 389 (1999)
61. N. Vranesevic et al., ApJ **617**, L139 (2004)
62. F.M. Walter, S.J. Wolk, R. Neuhauser, Nature **379**, 233 (1996)
63. M.C. Weisskopf et al., ApJ **652**, 387 (2006)
64. R. Wijnands, van der M. Klis, Nature **394**, 344 (1998)
65. P.M. Woods, C. Thompson, ed. by W.H.G. Lewin, M. Vander Klis in *Compact Stellar X-ray Sources* (CUP, Cambridge, 2004) (astro-ph/0406133)
66. D.G. Yakovlev, C.J. Pethick, Ann. Rev. Astr. Ap. **42**, 169 (2004)

Chapter 4
PMSingle: A Re-Analysis of the Parkes Multi-Beam Pulsar Survey in Search of RRATs

In this chapter, Sect. 4.3.2 is based on a paper published in the Monthly Notices of the Royal Astronomical Society, Eatough et al. [12] (astro-ph/0901.3993). Section 4.4 onwards is an enhanced version of a paper published in the Monthly Notices of the Royal Astronomical Society, [22] (astro-ph/0909.1924). As presented, this chapter describes the status of the search for RRATs in the PMSingle analysis as well as initial properties of the newly discovered sources, as they were in late 2009. Chapter 6 will give a completely up to date report on the timing solutions of these sources as well as additional confirmations since.

4.1 Observing Pulsars

Here we give a brief description of the steps involved in observations of pulsars using large radio telescopes. Observations are generally performed at sky frequencies of $\sim$1 GHz with bandwidths of a few 100 MHz. Observing frequencies of 3 GHz are considered 'high' [24] but surveys have been performed at 6.6 GHz [1, 45] and some pulsars have been observed at frequencies as high as 35 GHz [26, 28, 31]. On the 'low' frequency end, observations of pulsars are made routinely at the Puschino Observatory in Russia [37]. Commissioning observations with LOFAR have begun and observations with this instrument looks set to revolutionise low-frequency work [16, 47]. At the time of writing over 100 pulsars have been observed with LOFAR (T. Hassall, private communication).

Broadband pulsar emission is detected by a receiver of finite bandwidth, B, on a large radio telescope. The receiver is usually dual-polarisation and records either linear or circular polarisation. The incident electromagnetic radiation induces an electric current in the (conducting) antenna. The power received by, and current induced in, the antenna are related by $\langle P \rangle \propto I^2$, reminiscent of the heat dissipated by a resistor, $P = I^2 R$. The antenna is said to have a 'radiation resistance' and the output terminals of the antenna have a voltage across them, $V = IR$, which is

E. F. Keane, *The Transient Radio Sky*, Springer Theses,
DOI: 10.1007/978-3-642-19627-0_4, © Springer-Verlag Berlin Heidelberg 2011

what constitutes our detected raw signal. The voltage is directly proportional to the electric field[1] and we can form Stokes vectors from the signal.

Gigahertz frequencies are quite high for most telescope electronics to deal with and the frequency of the received voltage is usually mixed with a local oscillator to 'beat down' the frequencies, so that they span the frequency range $0-B$, where B denotes the bandwidth. The Nyquist-Shannon Sampling Theorem states that we must sample this input voltage at a rate of at least twice the bandwidth, the 'Nyquist frequency', $f_{\rm Nyq}$, in order to fully capture the information contained in the signal. For (say) a 500-MHz bandwidth we thus record two numbers every nano-second (one for each polarisation). This is what is commonly referred to as 'raw data' or 'baseband data'. There are now efforts being undertaken to coherently add baseband data between multiple telescopes [30], and some studies of nano-second features from the Crab pulsar have been made using baseband data [15], but usually it is converted to 'folded data' or 'search data'. In both cases the nanosecond time resolution is sacrificed in order to obtain frequency information. Taking data segments that are N samples in length and performing an FFT gives $N/2$ frequency channels for that segment. Doing this every N samples creates a 'filterbank'—a datacube of time, frequency and amplitude with time sampling of $N/2B = N/f_{\rm Nyq}$ and frequency resolution of $f_{\rm Nyq}/N$. At this point the data are further digitised to a desired dynamic range. Then, if the data are for timing analyses of a known pulsar, the effects of interstellar dispersion (described below) can be removed, following the FFT, by applying an appropriate transfer function, a process known as 'coherent dedispersion', which utilises the phase information still present in the data. In both cases the data are 'detected', i.e. squared to obtain intensity, rather than voltages. For folded data, the polarisation information is usually retained and the data will be averaged in phase according to a known ephemeris reducing the data into (say) 1-min sub-integrations. In the case of search data, as will be discussed below, the polarisations are typically summed to give total intensity. All of the data taken on RRAT sources are 'search mode' format.

4.2 Single Pulse Searches

Here we consider a time series of length T, an observation of a pulsar of period P. The number of periods is $N = T/P$. For a folded detection, i.e. averaged in pulse phase, the period-averaged flux density is given by the modified radiometer equation [35]:

$$S = \frac{G T_{\rm sys}}{\sqrt{n_{\rm p} B T}} \sqrt{\frac{\delta}{1-\delta}} (S/N). \tag{4.1}$$

[1] More precisely the amplitude of the frequency components of the voltage are proportional to the amplitudes of the corresponding electric field fluctuations.

Here G is the telescope gain (in Jy/K), T_{sys} is the system temperature, n_{p} is the number of polarisations summed to produce the time series, B is once more the bandwidth and δ is the pulse duty cycle. For detection of single pulses we consider a modified version of the form:

$$S_{\mathrm{peak}} = \frac{G T_{\mathrm{sys}}}{\sqrt{n_{\mathrm{p}} B W}} (S/N)_{\mathrm{peak}}, \tag{4.2}$$

where W is the pulse width. We can re-arrange this equation to give $(S/N)_{\mathrm{peak}} = \sqrt{n_{\mathrm{p}} B W} S_{\mathrm{peak}} S_{\mathrm{sys}}^{-1}$, if $S_{\mathrm{sys}} = G T_{\mathrm{sys}}$ is the noise equivalent flux density of the system. This is only the true peak density if the pulse is rectangular in shape; otherwise we actually measure ηS_{peak} where η is a shape-dependent constant of order unity. We note that the concept of a peak flux density is not well-defined when the peak is not well resolved, as it depends on the time sampling. If the time samples are a perfect match to the pulse width then we measure ηS_{peak}. An imperfect match reduces our signal-to-noise ratio by $(t_{\mathrm{samp}}/W)^{1/2}$ for $t_{\mathrm{samp}} < W$ or its inverse for $t_{\mathrm{samp}} > W$. It is common to assume this matching to be near perfect and we will proceed under this assumption. Practically this is achieved by having fast time sampling, i.e. a well-resolved pulse. We must also add a factor of β, where $1 - \beta$ is the fraction of the signal lost due to digitisation.

If we define r to be the ratio of single-pulse search and FFT search signal-to-noise ratios then we can determine an expression for r of the form:

$$r = \frac{A}{N^{1/2}} \frac{S_{\mathrm{peak}}}{S_{\mathrm{ave}}}, \tag{4.3}$$

where A is a product of constants of order unity, and is approximately equal to 2. Thus, for a pulsar to be more effectively detected in a single pulse search rather than an FFT search the strongest single pulse we see must be stronger than the average flux density by $\sqrt{N}$. For example, for an observation of 10^4 periods this means the peak flux density must be $\sim$100 times stronger than the average, a 'giant pulse' by the definition of Sect. 2.3.2. We can examine the behaviour of various flux density distributions (aka amplitude distributions) to determine the parameter space wherein $r > 1$, something which is useful to keep in mind for performing single pulse searches, especially with the hope of discovering sources entirely missed in FFT searches. If the pulses from the pulsar have a flux density distribution $f(S)$ then let $F(S)$, be the cumulative distribution

$$F(S) = \frac{\int_0^S f(S) dS}{\int_0^\infty f(S) dS}. \tag{4.4}$$

We defer the algebra to Appendix D, but it is straight-forward to show that:

$$F(S_{\mathrm{peak}}) = \frac{N-1}{N}. \tag{4.5}$$

We can use Eq. 4.5 to get S_{peak} for any $f(S)$ and the expectation value of observed pulses to obtain S_{ave} and hence determine r for any distribution.

Amplitude distributions for the 'normal' pulses from pulsars are generally seen to be lognormal ([2, 19, 46], private communication) whereas those for giant pulses are well described by power-laws [21, 36]. For illustration, we consider distributions which are lognormal, exponential, power-law and those with a nulling component. The latter three types of distribution were examined by McLaughlin and Cordes [41], whose results we agree with, with the exception of a few typos (see Appendix D).

(1) Lognormal:

$$f(S) = \frac{1}{\sqrt{2\pi}\sigma} \frac{e^{-\frac{(\ln S-\mu)^2}{2\sigma^2}}}{S} \tag{4.6}$$

The peak flux density is given by:

$$S_{\text{peak}} = \exp\left(\sqrt{2}\sigma \text{ erfinv}\left(1 - \frac{2}{N}\right) + \mu\right), \tag{4.7}$$

where erfinv denotes the inverse error function, and the measured average flux density is:

$$S_{\text{ave}} = \frac{e^{\mu+\frac{1}{2}\sigma^2}}{2}\left(1 + \text{erf}\left(\frac{(\ln S_{\text{peak}} - \mu) - \sigma^2}{\sqrt{2}\sigma}\right)\right). \tag{4.8}$$

We can see that as $S_{\text{peak}} \to \infty$, $S_{\text{ave}} \to \langle S \rangle = \exp\left(\mu + \frac{1}{2}\sigma^2\right)$, the expectation value of a lognormal distribution. In these expressions, S, S_{peak}, S_{ave}, μ and σ are all expressed in the same units. Cairns et al. [2] have determined $\mu = 2.3$ and $\sigma = 0.096$ for the Vela pulsar, and Weltevrede et al. [49, 50] have determined $\mu = -0.34$ and $\sigma = 0.99$ for the lognormal component of PSR B0656+14,[2] where in both cases μ and σ are in units of S_{ave}. The corresponding $r = r(N)$ curves for these values are plotted in Fig. 4.1. We can see that pulsars with the Vela-like distributions in the 35-minute PMPS observations are more significantly detected in single pulse searches if $P > 4.5$ s, while, for the B0656-like distributions, single pulse searches in the PMPS are only superior for $P > 7.5$ s. Note that given the periods of Vela and B0656+14 (89.3 ms and 384.9 ms respectively) they would both be more easily detected in the PMPS in an FFT search. The actual periods of the two pulsars are marked with arrows in Fig. 4.1.

(2) Exponential:

$$f(S) = \frac{1}{\sigma} e^{-\frac{S}{\sigma}} \tag{4.9}$$

[2] It is important to note that the authors state that this pulsar's distribution seems to be multi-modal, i.e. a lognormal alone is not sufficient to describe $f(S)$, but we use their measured values for illustration.

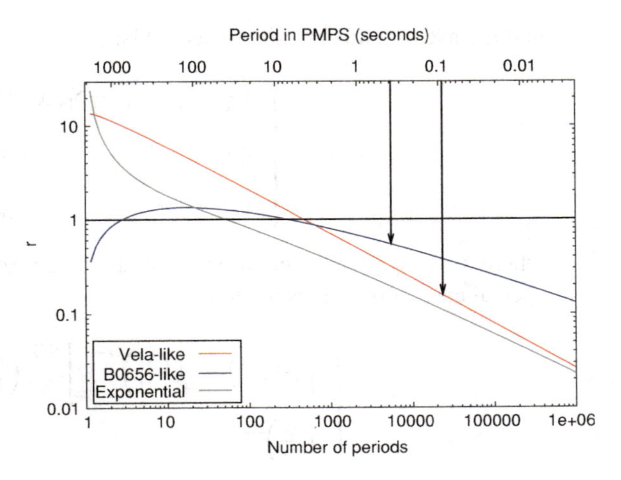

Fig. 4.1 Plotted is $r = r(N)$ for two different lognormal distributions (*red* and *blue curves*), and the exponential distribution (*grey curve*). Alternatively, r can be read as a function of the corresponding period in the PMPS, on the upper horizontal scale. Points on the curves above the $r = 1$ line imply the pulsar will be more significantly detected in a search for single pulses. Below this line FFT searches are more effective

Equation 4.5 immediately yields $S_{\text{peak}} = \sigma \ln N$. The average and peak flux densities are related by:

$$S_{\text{ave}} = S_{\text{peak}} \left(\frac{N-1}{N}\right)\left(\frac{1-(\ln N)/(N-1)}{\ln N}\right), \tag{4.10}$$

so that :

$$r = A\frac{N^{\frac{1}{2}}}{N-1}\left(\frac{1-(\ln N)/(N-1)}{\ln N}\right). \tag{4.11}$$

which is independent of any parameters. Figure 4.1 also shows $r = r(N)$ for the exponential distribution. We can see that single pulse searches are superior only while $N < 47$, i.e. periods greater than 44 s in the PMPS.
Power-laws:

$$f(S) = AS^{-\alpha} \tag{4.12}$$

Assuming the pulses are emitted in a finite range, between S_1 and S_2 (so that $f(S)$ does not diverge), we can calculate the peak flux density observed, to be:

$$S_{\text{peak}} = \begin{cases} \left[\left(\frac{N-1}{N}\right)S_2^{1-\alpha} + \frac{1}{N}S_1^{1-\alpha}\right]^{\frac{1}{1-\alpha}} & \alpha \neq 1 \\ S_1\left(\frac{S_2}{S_1}\right)^{\frac{N-1}{N}} & \alpha = 1 \end{cases} \tag{4.13}$$

and the average flux density observed is:

$$S_{\text{ave}} = \begin{cases} \left(\frac{\alpha-1}{\alpha-2}\right)\frac{x^{\alpha-2}-1}{x^{\alpha-1}-1} S_{\text{peak}} & \alpha \neq 1,\ 2 \\ \frac{(S_{\text{peak}} - S_1)}{\ln(S_{\text{peak}}/S_1)} & \alpha = 1 \\ \frac{S_{\text{peak}} S_1 \ln x}{S - S_1} & \alpha = 2 \end{cases} \tag{4.14}$$

where $x = S_{\text{peak}}/S_1$, which, from re-arranging Eq. 4.13, and evaluating the special case of $\alpha = 1$, is given by:

$$x = \begin{cases} \left[\left(\frac{N-1}{N}\right) + \frac{1}{N}\left(\frac{S_2}{S_1}\right)^{\alpha-1}\right]^{\frac{1}{1-\alpha}} \left(\frac{S_2}{S_1}\right) & \alpha \neq 1 \\ \left(\frac{S_2}{S_1}\right)^{\frac{N-1}{N}} & \alpha = 1 \end{cases} \tag{4.15}$$

The expression for $r = r(N)$ is thus:

$$r = \begin{cases} \frac{A}{N^{\frac{1}{2}}}\left(\frac{\alpha-2}{\alpha-1}\right)\frac{x^{\alpha-1}-1}{x^{\alpha-2}-1} & \alpha \neq 1,\ 2 \\ \frac{A}{N^{\frac{1}{2}}} \ln x \frac{x}{x-1} & \alpha = 1 \\ \frac{A}{N^{\frac{1}{2}}} \frac{x-1}{\ln x} & \alpha = 2 \end{cases} \tag{4.16}$$

If we choose a value for S_2/S_1 we can examine the behaviour of $r = r(N)$ for a range of power-law indices, as shown in Fig. 4.2. We can see that the trend for large N is a $\sqrt{N}$ decrease, i.e. as the length of your observations increase the FFT signal-to-noise ratio will always, eventually, surpass the single pulse signal-to-noise ratio. For low values of N it is always easier to detect something via its single pulses. However, the curves are not simply decreasing for all N and for $0 < \alpha < 3$, $dr/dN = 0$ before $N \to \infty$, i.e. there is a 'sweet spot', a peak in $r = r(N)$ where single pulse searches can be more effective than FFT searches with detections an order of magnitude stronger. We note that only those power laws with $1 < \alpha < 3$, have peaks after $N = 2$, so that effectively these are the power laws where single pulse searches can be most effective. Figure 4.2 shows power-law curves for $\log(S_2/S_1)$ ranging from 2 to 5. These seem to be realistic ranges given that Karuppusamy et al. [21] observed $S_{\text{peak}}/S_{\text{ave}} = 7 \times 10^5$ in the Crab, Cognard et al. [6] observed $S_{\text{peak}}/S_{\text{ave}} = 4.2 \times 10^3$ in PSR 1937+21 and Welteverede et al. [49, 50] observed $S_{\text{peak}}/S_{\text{ave}} = 116$ in PSR B0656+14.

(4) Bimodal/Nulling:

$$f(S) = (1-g)\delta(S - S_1) + g\delta(S - S_2) \tag{4.17}$$

Assuming there is at least one detectable pulse of flux density S_2, then $gN \geq 1$ and, clearly, the average flux density is just $(1-g)S_1 + gS_2$. The peak flux is S_2 so that:

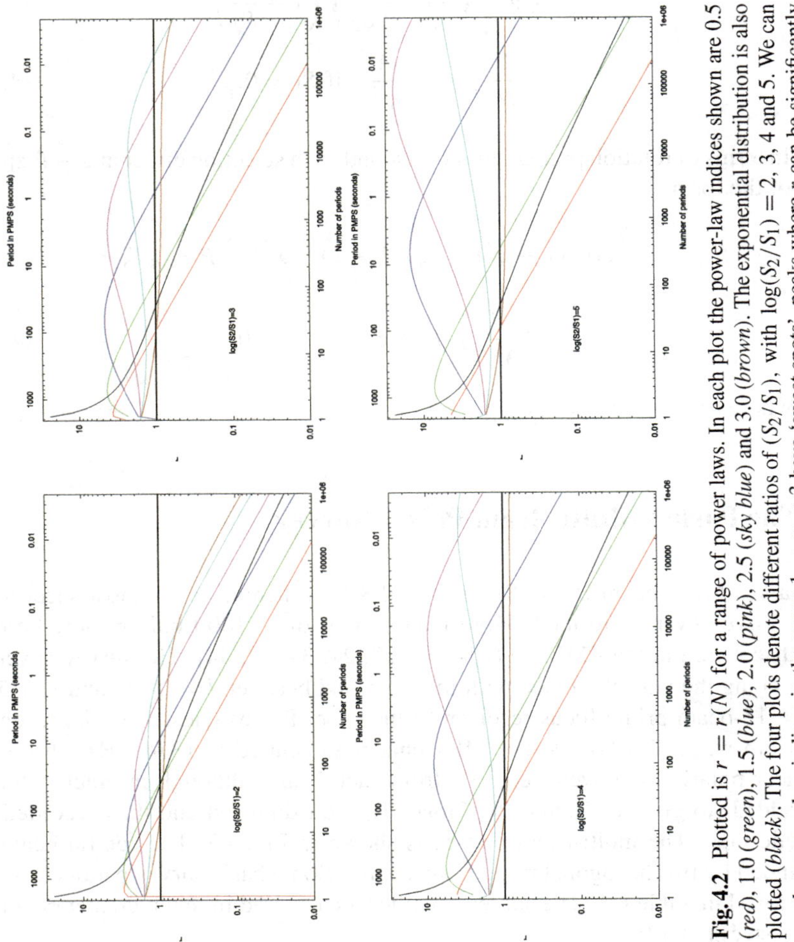

Fig. 4.2 Plotted is $r = r(N)$ for a range of power laws. In each plot the power-law indices shown are 0.5 (*red*), 1.0 (*green*), 1.5 (*blue*), 2.0 (*pink*), 2.5 (*sky blue*) and 3.0 (*brown*). The exponential distribution is also plotted (*black*). The four plots denote different ratios of (S_2/S_1), with $\log(S_2/S_1) = 2, 3, 4$ and 5. We can see that power-law indices in the range $1 < \alpha < 3$ have 'sweet spots', peaks where r can be significantly higher than 1 for a wide range of N. The upper horizontal scale gives the corresponding PMPS period

$$r = \frac{A}{N^{\frac{1}{2}}} \frac{S_2}{(1-g)S_1 + gS_2}. \tag{4.18}$$

There are regions of $g - N$ space where $r \geq 1$ which depend on (S_1/S_2), unless of course $S_1 = 0$, i.e. a nulling component. These regions are defined as:

$$\frac{1}{N} \leq g \leq \left(\frac{A}{N^{\frac{1}{2}}} - \frac{S_1}{S_2}\right)\left(1 - \frac{S_1}{S_2}\right)^{-1}, \tag{4.19}$$

$$\frac{1}{N} \leq g \leq \frac{A}{N^{\frac{1}{2}}} \quad \text{if } S_1 = 0. \tag{4.20}$$

In terms of rotation period, this corresponds to a selection effect in $g - P$ space of the form:

$$\frac{T}{A^2}[g^2(1-x^2) + 2gx(1-x) + x^2] \leq P \leq gT, \tag{4.21}$$

$$\frac{g^2 T}{A^2} \leq P \leq gT \quad \text{if } S_1 = 0, \tag{4.22}$$

where once again $x = S_{\text{peak}}/S_1$.

4.3 The Parkes Multi-Beam Pulsar Survey

The Parkes Multi-beam Pulsar Survey (PMPS from herein) is the most successful pulsar survey ever performed. More than half of the ~1800 pulsars now known were discovered in the PMPS [11, 14, 17, 23, 27, 34, 38, 44]. The survey covered a strip along the Galactic plane with $|b| < 5°$ and between $l = 260°$ and $l = 50°$ using a 13-beam prime-focus receiver on the 64-m Parkes telescope. The receiver bandwidth was 288 MHz (96 × 3 MHz channels) centred at 1374 MHz. Orthogonal linear polarisations were received from each beam, although in practise these were added (to give total intensity, Stokes I), 1-bit digitised and then recorded to magnetic tape. The multi-beam pattern is shown in Fig. 4.3. The central beam is surrounded by two hexagonal rings of six beams. Some basic survey parameters are summarised in Table 4.1. The survey specifics are covered in much greater detail in Manchester et al. [38].

4.3.1 *Why Re-process?*

By mid 2008, a re-processing of the PMPS was timely for a number of reasons: (1) A new interference removal algorithm known as the 'zero-DM filter' had been

Table 4.1 The survey parameters for the PMPS [38]

Parameter	Value
Galactic latitude	$\|b\| < 5°$
Galactic longitude	$260° - 50°$
Central frequency	1374 MHz
Bandwidth	96×3 MHz
Sampling time	250 μs
Pointing length	35 min = 2^{23} samples
Number of pointings (beams)	3167 (41561)
Polarisations/beam	2
Telescope gain (K/Jy)	0.735 (central beam)
	0.690 (inner ring)
	0.581 (outer ring)
Half-power beamwidth (arcmin)	14.0 (central beam)
	14.1 (inner ring)
	14.5 (outer ring)
Beam ellipticity	0.0 (central beam)
	0.03 (inner ring)
	0.06 (outer ring)

Table 4.2 HYDRA specifications

Parameter	Value
No. of nodes	181
Processors/node	$2 \times$ Quad Core Intel Xeon 2.66 GHz E5430 CPU
Memory/node	4 GB RAM
Disk space/node	2×250 GB Disks
Entire survey processing time	~9.5 years (1 node)
Real time using DCORE	~4 months (15/36 nodes)
Real time using HYDRA	~2 weeks (60/181 nodes)

developed, which we describe below; (2) The discovery of RRATs, and their implied population size meant that there was an expectation of more sources yet to be discovered in the PMPS. The RFI excision possible with the zero-DM filter would open up the possibility of new discoveries in data which had previously been riddled with RFI; (3) The Jodrell Bank pulsar group had just acquired the HYDRA supercomputer—a 1448 processor cluster which constituted a significant upgrade from the 72-processor DCORE cluster which had been in use to that point. Such computing abilities would reduce the time it would take to process the entire survey so that it would be feasible in a few months. Previous analyses had taken well over a year. The specifications of HYDRA are given in Table 4.2; (4) Finally, we can improve our search with some simple additions: we can search for a wider range of pulse widths, perform a beam comparison to remove 'multi-beam events' for extra RFI excision, utilise GUIs for easier examination of candidates and we can improve the survey book-keeping.

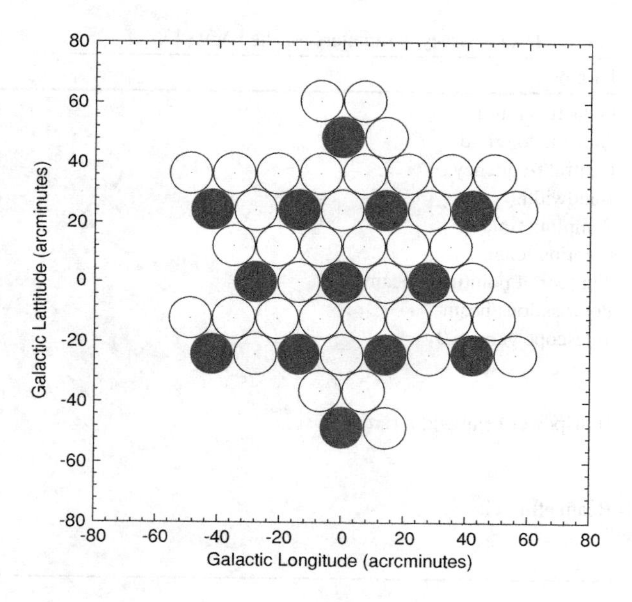

Fig. 4.3 The beam pattern of the PMPS multi-beam receiver. The 13 beams observed a region of sky, such as the regions with the shaded beams. To fill in the gaps in sky coverage the next three pointings shift by one half-power beamwidth ($\sim$14 arcmin), as shown. This tessellation of beams is repeated across the entire survey region

4.3.2 The Zero-DM Filter

Pulsar search data are often corrupted by the presence of impulsive, broadband and sometimes periodic terrestrial radiation—radio frequency interference (RFI). This radiation originates in unshielded electrical equipment which produces discharges, such as motor car ignition systems, electric motors and fluorescent lighting, as well as from mobile phone signals, television transmissions, radar and lightning (a particular problem at Parkes) to name but a few. This can often be very strong and enters receiver systems through the far sidelobes of the telescope reception pattern. Several methods have been employed to reduce the effects of this radiation, such as clipping the intense spikes, filtering parts of the fluctuation spectrum, and identifying common signals in the different beams of a multi-beam receiver system. These procedures all require carefully tuned algorithms to remove the interference, while at the same time causing minimal damage to the astronomical data. These impulsive RFI signals are: (1) mostly broadband; and (2) generally do not display the dispersed signature of radio pulsars. We have developed a simple algorithm which we call 'zero DM filtering' which makes use of these two properties to selectively remove broadband undispersed signals from the data prior to the application of normal pulsar search algorithms [12].

Broadband pulsar signals arrive at Earth dispersed due to their journey through the ISM—a dispersive medium with a frequency dependent refractive index. Pulses are delayed with respect to the light travel time L/c by an amount t in a frequency dependent way which is:

$$t = 4150\left(\frac{\mathrm{DM}}{f_{\mathrm{MHz}}^2}\right) \ \mathrm{sec} \tag{4.23}$$

where DM, the dispersion measure of a source at a distance L from Earth, is:

$$\mathrm{DM} = \int_0^L n_{\mathrm{e}}(l)dl, \tag{4.24}$$

where $n_{\mathrm{e}}(l)$ is the electron density at a distance l. DM is conventionally measured in units of cm^{-3} pc. Detecting a pulse with a finite bandwidth receiver results in a broadened pulse profile and a corresponding reduction in pulse signal-to-noise ratio. To compensate for the effects of this dispersion, before any pulsar search can be performed, the data are 'de-dispersed' at a set of trial DM's. In order to do this, the bandpass of the receiver is first split into a number of independent frequency channels (a 'filterbank') as described in Sect. 4.1 Appropriate time delays for a given trial DM are then applied to each frequency channel so that the pulses arrive at the output at the same time. The channels are then summed together to produce a 'dedispersed' time sequence. A blind pulsar search will typically do this for a large number of trial DMs. Pulsar signals will peak at some $\mathrm{DM} > 0$ but as RFI is terrestrial in nature it peaks at $\mathrm{DM} = 0$. The zero-DM filter exploits this fact to remove such non-dispersed signals. The filter is implemented by simply calculating the mean of all frequency channels in each time sample and subtracting this from each individual frequency channel in the time sample. For a given time sample, t, the correction to the value recorded at frequency channel f_{i} is:

$$S(f_{\mathrm{i}}, t) \longrightarrow S(f_{\mathrm{i}}, t) - \frac{1}{n_{\mathrm{chans}}} \sum_{j=1}^{n_{\mathrm{chans}}} S(f_{\mathrm{j}}, t). \tag{4.25}$$

After applying this filter we dedisperse and proceed as normal. We consider an idealised dispersed pulsar signal with very narrow pulses and approximate the dispersion drift across the full filterbank bandwidth B to a linear slope, df/dt, which is given approximately by:

$$\frac{dt}{df} = -8300\left(\frac{\mathrm{DM}}{f_0{}^3}\right) \ \mathrm{s/MHz}, \tag{4.26}$$

where f_0 (MHz) is the central observing frequency. Figure 4.4a shows such a pulse sweeping through the filterbank channels of a receiver, traversing the full bandwidth B in a time of Bdt/df. After subtraction of the mean in vertical strips as described above, the non-pulse grey area in Fig. 4.4a is negative. Dedispersion then distorts this pattern in the manner shown in Fig. 4.4b. Adding vertically after the dedispersion then gives the dedispersed time-sequence shown in Fig. 4.4c, in which the negative area of the triangle equals the area of the pulse.

Thus the dedispersion process results in the data being convolved by this function, $w(t)$, or equivalently the fluctuation spectrum being multiplied by its Fourier transform, $W(\nu)$, which is:

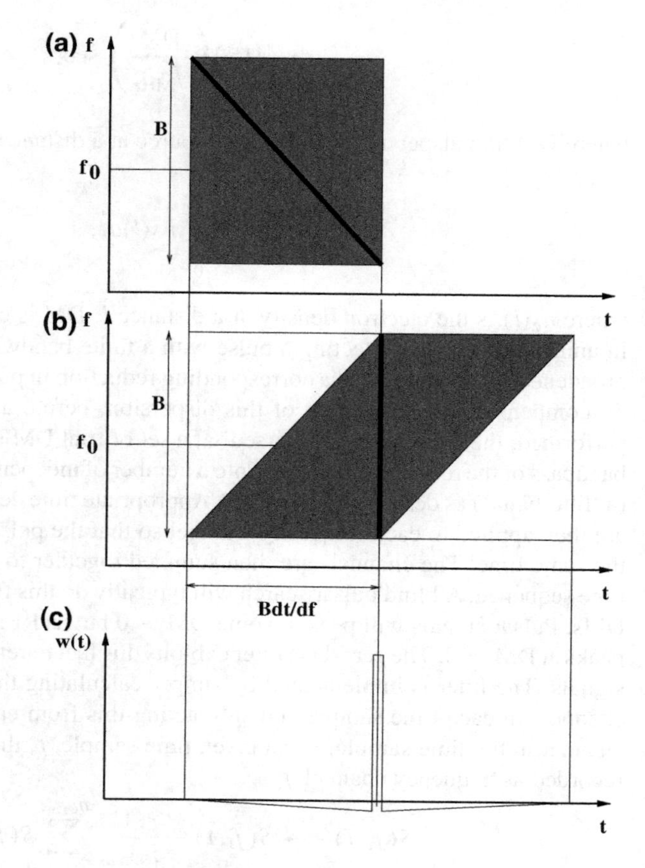

Fig. 4.4 Panel **a** shows the dispersion drift of the pulse *Bdt/df* over the full bandwidth B of the receiver. Panel **b** shows the result of dedispersing the data at the correct value of DM and Panel **c** shows the resultant pulse shape after adding all the frequency channels together

$$
\begin{aligned}
W(\nu) &= 1 - \mathrm{sinc}^2\left(\pi B \frac{dt}{df} \cdot \nu\right), \\
&= 1 - \mathrm{sinc}^2\left(\frac{8300\pi B \cdot DM}{f_0^3} \cdot \nu\right),
\end{aligned}
\tag{4.27}
$$

where ν is the fluctuation frequency of the signal. We re-iterate that f denotes observing/sky frequency, while ν is used to denote the fluctuation frequencies within the data. Figure 4.5 shows the effective filters applied for two specified DMs for PMPS observations. The half-amplitude point of the functions is at $\sim$500/DM Hz. Thus for DM $= 100\,\mathrm{cm}^{-3}$ pc, all frequencies <5 Hz will be lost, i.e. Periods >200 ms. Of course, a normal pulsar with such a period will have most of its power in harmonics which will not be affected at all, especially since standard pulsar search techniques perform harmonic summing [35].

There is also a second step to the filtering process where we consider the effects of $w(t)$ convolution in searching the time series for single pulses. Standard single pulse searches involve convolving a de-dispersed time series $x(t)$ with box-cars $b(t)$ of

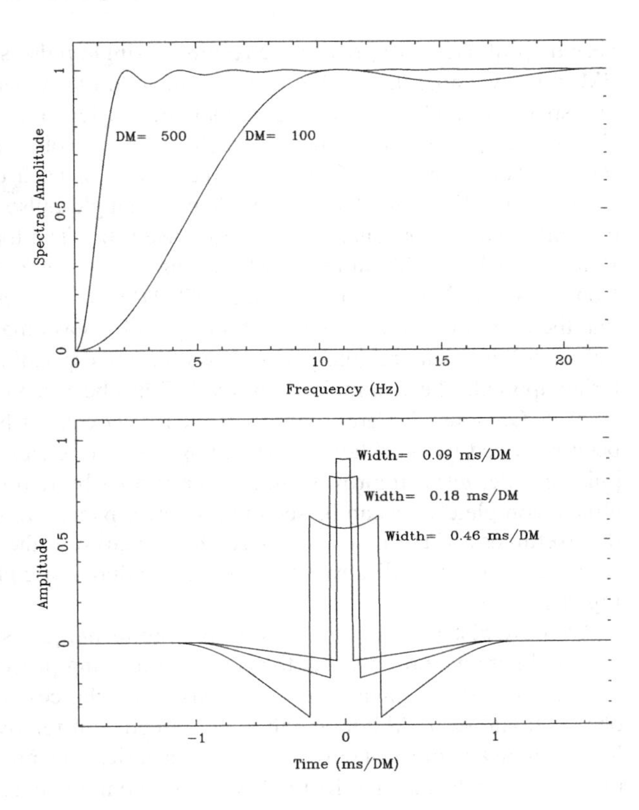

Fig. 4.5 *Top*: Two example zero-DM filters for DM $= 500\,\mathrm{cm}^{-3}$ pc and DM $= 100\,\mathrm{cm}^{-3}$ pc for the PMPS, showing variation in relative spectral amplitude with the fluctuation frequency. *Bottom*: The effect of zero-DM filtering on square pulses in the PMPS. The time-scale should be multiplied by the trial value of DM to give time in milliseconds

various widths and searching for peaks in $x(t) * b(t)$ where $*$ denotes convolution. In removing non-dispersed RFI we have also convolved individual pulses with $w(t)$ which reduces their amplitudes as shown in Fig. 4.5. Due to the removal of RFI, the overall effect can be to make such single pulses more visible in single pulse search diagnostic plots. However, this amplitude reduction can be recovered by deconvolving the time series with $w(t)$. The zero-DM filtered time series is now $x'(t) = x(t) * w(t)$ so that a standard single pulse search looks for peaks in $x'(t) * b(t)$. As a result of zero-DM filtering the optimal filter is now not a box-car but rather is given by $b'(t) = b(t) * w(t)$. We now search for peaks in $x''(t) = x'(t) * b'(t) = F^{-1}[X'(\nu) * W(\nu)] * b(t)$ where we have used the commutability of convolution so that we can perform the convolution in the frequency domain which in practise is what we do. This is desirable as convolution is simply a multiplication in the frequency domain. Two advantages of this are that the fluctuation spectrum has had the optimal filter $W(\nu)$ applied to it before performing a periodicity search, and, as single pulse searches are normally implemented in tandem with periodicity searches there is just one extra FFT required to search for single pulses in the time domain.

We apply the zero-DM filter having removed all clipping and zapping algorithms used previously for the mitigation of RFI [17]. The effects are as expected and very helpful. Here we present the results of single pulse searches as applied to the PMPS but we note that FFT searches benefit similarly with the filtering producing less spurious candidates making real sources easier to identify [11–13]. Figure 4.6 shows examples of single pulse search diagnostic plots from our analysis of the PMPS. Plotted in Fig. 4.6. are the time series for each of the 325 trial DMs (in the range 0–2200 cm^{-3} pc). Each detected single-pulse event is plotted as a circle with area proportional to signal-to-noise ratio. The top panel is the result of a standard analysis without zero-DM filtering. The observation contains single pulses from PSR B1735−32 at DM channel 82 (DM $=$ 49.9 cm^{-3} pc), but we can see that the output is contaminated with many RFI streaks making identification of the pulsar difficult. The middle panel shows the corresponding plot with zero-DM filtering applied. The vast majority of the RFI has been removed but the pulsar is still present. Obvious also are a few remnant RFI streaks at high DM which have not been removed by the filtering. Searching the time series with the optimal single-pulse profile, $w(t)$, improves things even further by removing the remnant streaks almost completely, as can be seen in the lower panel. Another example illustrating the usefulness of the new filter is given by analysing the observation wherein the 'Lorimer Burst' was discovered. Single pulse diagnostic plots for this are shown in Fig. 4.7.

We have discussed the dangers of filtering techniques such as clipping and zapping and here we propose that the zero-DM filtering process is a more natural way of removing RFI signals free of any arbitrary choices of where and how to cut observational data. Although extremely effective at removing the vast majority of RFI we note a number of caveats and practicalities one must be aware of when using this new filter. Broadband RFI at DM $= 0$ is completely removed but the filter also reduces the signal-to-noise ratio of real signals with a low dispersion (see Fig. 4.5). This means that the technique will not be of much benefit to high-frequency, or narrow-bandwidth searches because in this case even signals from celestial sources will not be highly dispersed (see Eq. 4.23). For example, the dispersive delay across the entire 576-MHz bandwidth in the Parkes Methanol Multi-Beam Survey (centred at 6.59 GHz) is less than 2 ms for a source with DM $= 100\,\text{cm}^{-3}$ pc, giving a high-pass cut-off frequency of 154 Hz, making detection of all but MSPs impossible. As MSPs are weak they are more easily detectable nearby, implying low dispersion, so that the MMB survey is insensitive to them [1] meaning that using the zero-DM filter would essentially remove all the pulsars from this survey. Another point of note is our assumption of a linear dispersion slope when the true slope is in fact quadratic. A complete description of the dispersion slope increases the difficulty of implementing the algorithm with little benefit. The effects of this assumption manifest themselves in phase-folded pulse profiles which have asymmetric dips either side of the pulse unlike the symmetric dips which we have considered. An extra cautionary note is required in single-pulse searches where RFI signals can occasionally be strong enough to persist even after zero-DM filtering. In this case entire RFI streaks are not removed resulting in remnant streaks, i.e. high signal-to-noise peaks at non-zero DM. However such remnant RFI streaks will not have the expected shape as a function of trial DM, nor will they appear dispersed according to

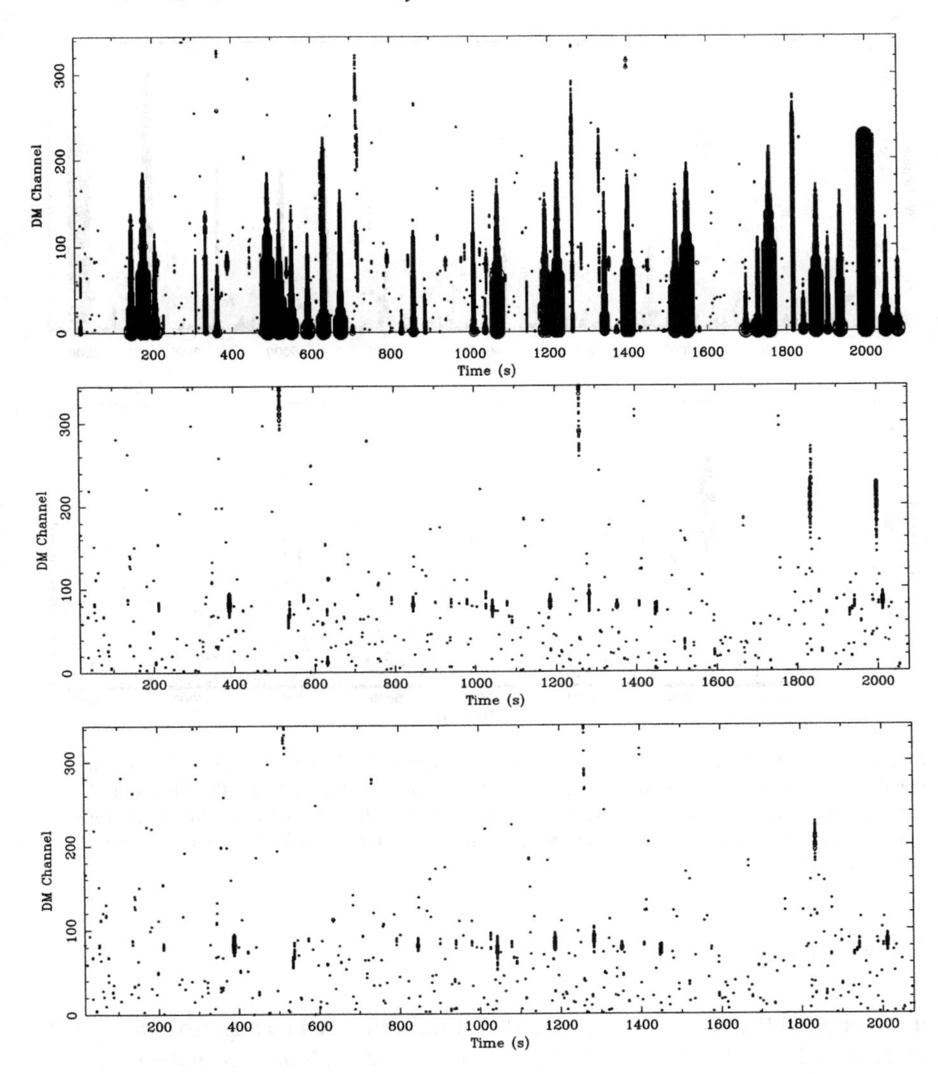

Fig. 4.6 Single pulse search diagnostic plots. *Top*: using standard search method without zero-DM filter; *Middle*: After using zero-DM filter most RFI streaks have been removed with single pulses from PSR B1735−32 (at DM channel 82) now much more visible; *Bottom*: After using the filter and applying the optimal matched filter. Remnant RFI streaks are further removed and pulses from the pulsar remain

Eq. 4.23, like a true celestial source. They are also likely to appear in multiple beams of a multi-beam receiver. Each of these properties can be used to further remove such spurious sources.

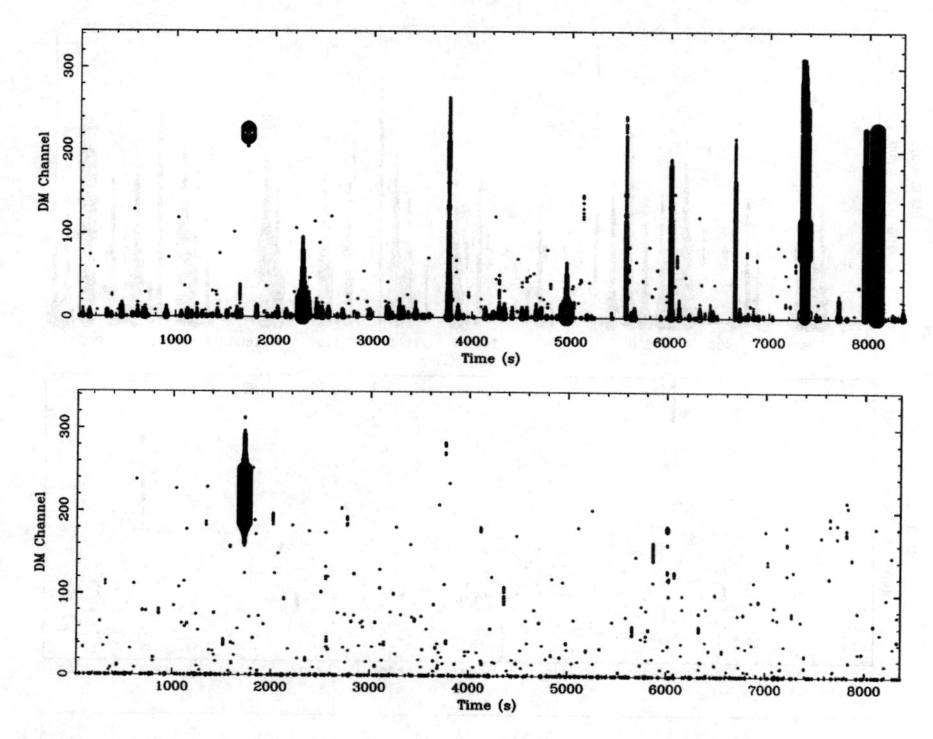

Fig. 4.7 *Top*: The burst is seen at ~1750 s at DM channel ≈220 with a $S/N < 20$. Much RFI streaks peaking at DM = 0 are visible. *Bottom*: The same plot after using the filter (and beam comparisons—see Sect. 4.4 for discussion). We can see that the RFI is almost completely removed and the burst detection is much more secure now with $S/N = 37$. Data kindly provided by D. R. Lorimer

4.4 PMSingle

In the original RRAT discovery paper [43], it was estimated that approximately half of the RRATs visible in the PMPS had been detected, with the remainder obscured due to the effects of impulsive RFI. Recently it has become timely to re-process the PMPS survey in search of these postulated sources as we have developed a new and effective RFI mitigation scheme ([11, 12]; Sect. 4.3.2). Utilising the Jodrell Bank Pulsar Group's recently acquired 1448-processor HYDRA super-computer, the entire PMPS data set was re-processed. Modified versions of the *Sigproc*[3] processing tools were used. In the following we refer to this project as the PMSingle analysis. Supplementary book-keeping information for this analysis is available online.[4]

[3] http://sigproc.sourceforge.net/

[4] http://www.jb.man.ac.uk/~ekean/pmsingle/pmsingle.htm.

We note that the presence of impulsive RFI will make a search somewhat blind to sources with low dispersion measure (DM). RFI is terrestrial in origin and, not having traversed the inter-stellar medium, we expect it to have $DM = 0$.[5] However as RFI signals are typically very strong, in comparison to the relatively weak astrophysical signals of interest, they are seen with sufficient residual intensity to mask celestial signals, often to high DM values. Figure 4.8 shows an example of this 'low-DM blindness' due to the presence of strong terrestrial RFI. This of course means that searches can miss real, low-DM celestial sources. Thus our sensitivity to the nearby Galactic volume may have been reduced due to the effects of RFI in the initial analysis.

Our PMSingle processing involved the following steps:

1. *Remove all clipping algorithms.* In previous analyses of the PMPS the raw data have been 'clipped', i.e. the data were read in 48 KB blocks and those blocks wherein the sum of all the bit values was larger than some (user supplied) threshold above the mean had their values set to the mean level (half 1s and half 0 s in the case of 1-bit data). The motivation for such clipping is that RFI signals are typically much stronger than real astrophysical signals, so that the brightest detections are taken to be RFI spikes. This, however, is not optimal in that it removes signals based on strength and the discovery of RRATs and pulsars which show strong single pulses show that such signals may be of astrophysical interest. The threshold used is also arbitrary and usually determined on a trial-and-error basis. We note that the strongest pulses with high dispersion (i.e. dispersed over 2 or more blocks) can escape being clipped in error, so that low-DM sources are the most likely to be clipped in this way. We therefore removed this step from our re-processing.
2. *Dedisperse the raw filterbank data using the zero-DM filter.* We searched for dispersed signals in a DM range of 0–2200 cm^{-3} pc. For the Galactic longitudes covered by the PMPS this corresponds, at $|b| = 0°$, to typical distances of $\gtrsim$40 kpc at $l = 260°$ and $l = 50°$ to 8.5 kpc towards the Galactic centre.
3. *Search for bright single pulses.* This is an exercise in matched filtering using box-cars of various widths in time in the dedispersed time sequences. However, instead of rectangular box-cars the zero-DM filtering means it is optimal to search with box-cars which have been convolved with the zero-DM filter function. The effect of the zero-DM filter on a pulse from the newly discovered source, J1841−14 are shown in Fig. 4.9. In PMSingle, at the lowest DMs, we search for pulse widths as narrow as 250 μs to as wide as 128 ms. As we increase the DM trial value we get dispersive smearing of pulses to widths much longer than their intrinsic widths so at these DMs we search for even wider pulses, i.e. from 500 μs–256 ms, then 1–512 ms, etc. The widest pulse widths searched for are a factor of 16 larger than in the original PMPS single-pulse search analysis [40, 43].

[5] RFI signals emerging from air-traffic control radar, a particular problem at frequencies near 1400 MHz, are sometimes observed to also show signals sweeping in frequency. We discuss some of the effects of radar signals in Sect. 5.2.4.

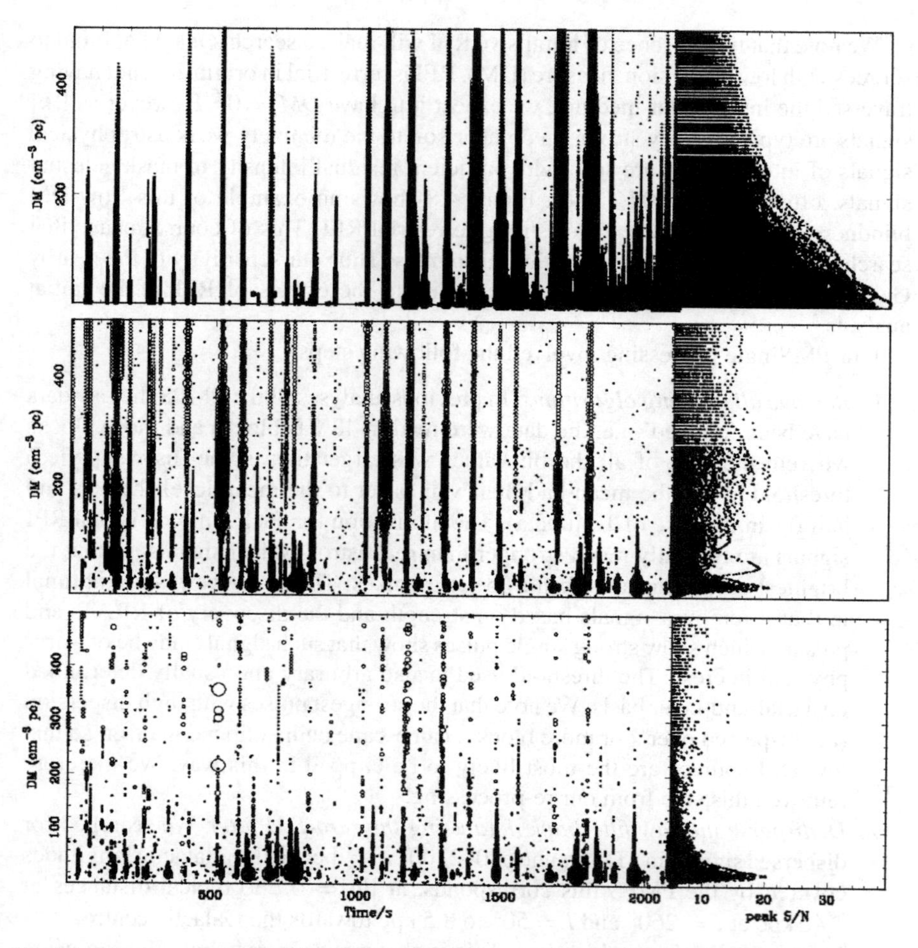

Fig. 4.8 DM-time plots for a 35-min PMPS observation. The ordinate in the large left-hand plots is trial DM over the range 0–500 cm^{-3} pc and the abscissa is time. To the right are plots of trial DM versus peak signal-to-noise ratio. Significant detected events are plotted as circles with radius proportional to signal-to-noise. **a** The strong vertical stripes across wide DM ranges are instances of extremely strong RFI. Inspection of the plot for the presence of a celestial source is impossible due to the presence of this RFI. **b** The same beam after the zero-DM filter has been applied (see Sect. 4.3.2). We can see that the RFI has not been completely removed, especially at higher DMs. However the RFI has been removed much more effectively at low DMs and a source at DM$\sim$20 cm^{-3} pc is beginning to become visible. **c** The same beam after application of the zero-DM filter and the removal of multiple-beam events. We can see that the diagnostic plot is cleaned up even further and, although there is still remnant RFI, it is clear that there is a real source in this beam at DM $\sim$ 20 cm^{-3} pc. This is the first detection of J1841$-$14 in the PMPS, the lowest DM source found so far in the survey

4. *Perform a beam comparison to remove multi-beam events.* The zero-DM filtering is effective at removing short-duration broad-band RFI. The more persistent or narrow-band that impulsive RFI is, the less likely it will be completely

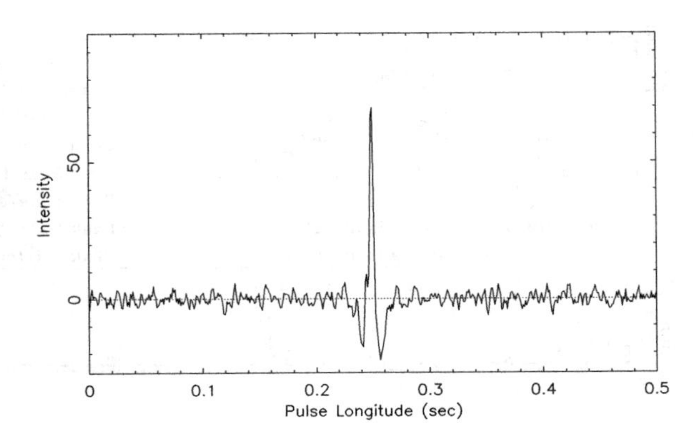

Fig. 4.9 An example of a single pulse detected from J1841−14. Here we plot 0.5 s of the 6.6 s period centred on the pulse with intensity plotted in arbitrary units. Evident are the dips either side of the pulse which is a remnant of the zero-DM filtering process

removed. However as the PMPS used a 13-beam receiver we have extra information to help with RFI mitigation. Pulsar signals are very weak and typically are seen in only one beam. The strongest pulsars (e.g. Vela) can be seen in a few beams but normally no more than three. Even the extremely bright 30-Jy 5-ms Lorimer Burst was seen in just three beams [33]. We can thus apply a rejection criterion for detected events like: for each detected event—(DM, time) point, if we have detections in the range (DM $\pm\, \varepsilon_{\rm DM}$, time $\pm\, \varepsilon_{\rm time}$) in, say, $\geq$5 beams in that pointing then ignore this detection as it is most likely RFI. We conservatively took ε to be one bin in each case (i.e. one DM trial step and one time sample step).

5. *Produce diagnostic plots for inspection and classification.* Along the lines of Cordes & McLaughlin [9] a series of diagnostic plots are created for each beam. An example of this is shown in Fig. 4.10. The plots include beam information (beam and pointing number, sky position etc.) as well as information on the number of multiple beam detections which were removed. Each beam was inspected and classified as containing either noise, a known pulsar, a known RRAT or a new candidate—divided into Classes 1, 2 and 3. Examples of each of these classes are given in Fig. 4.11. Class 1 candidates are all judged to be real sources, either yet-to-be confirmed RRATs or known pulsars detected in the telescope's far-side-lobes. Class 3 candidates are weak and experience suggest that no confirmations are expected. Class 2 sources are intermediate between these classes. Beams could also be classified as being too adversely affected with remnant impulsive RFI (not removed by zero-DM filtering or beam comparisons) so as to make inspection impossible. In these beams a real source, unless it were very strong, would not have been detectable. The results of the classifications are given in Table 4.3.
6. *Cross-check with known sources.* For each candidate we confirm that it is not a known pulsar (or known RRAT). Even if there is no pulsar within the telescope beam the pulses could still be from a known (strong) pulsar perhaps several beamwidths away on the sky (e.g. PSRs B0835–41, B0833–45 and B1601–52 are detected many times like this). To do this we can compare

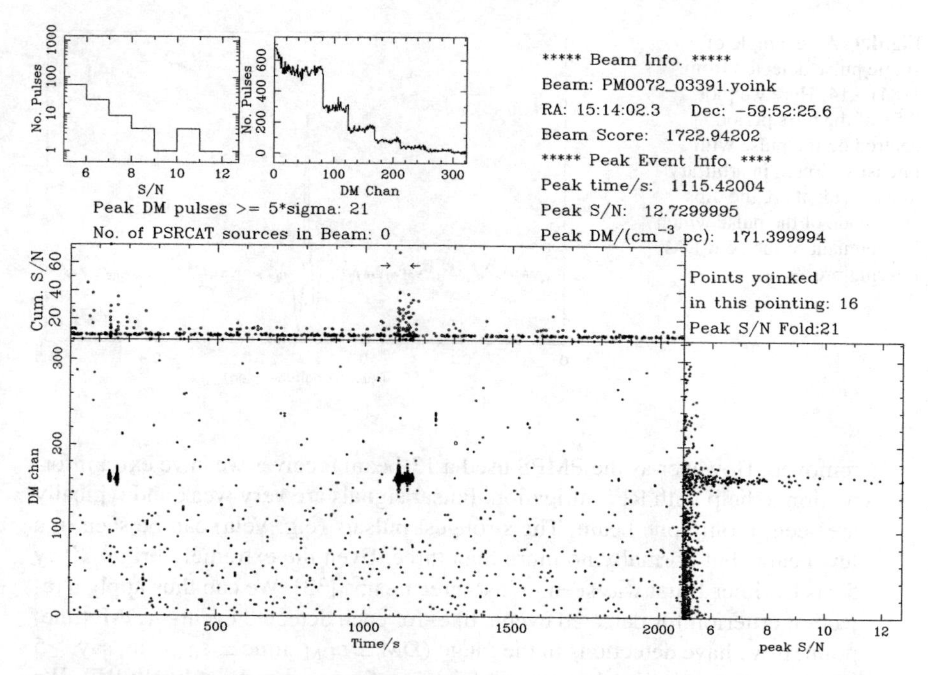

Fig. 4.10 An example of the diagnostic plots. This plot shows the original detection of J1514−59. Two clumps of pulses are evident at ~150 s and at ~1100 s. The steps evident in the second upper histogram, for the number of pulses as a function of DM, is a result of down-sampling the data by a factor of two upon reaching the 'diagonal DM', where adjacent time samples are no longer independent [35]

the position (with a larger tolerance) and DM of the candidate to the value for known sources.

7. *Determine a period for the candidate.* As RRATs are not seen in FFT searches we use a method of factorising the pulse time of arrival (TOA) differences to determine the period. For N TOAs there are at most[6] $(N/2)(N-1)$ unique TOA differences which we can use. For some small increment we step through a large range of trial periods. At the correct period, an integral number of periods will fit all of the TOA differences with a small error and rms residual. Harmonics of the true period will also match many of the TOA differences but at a less significant level. If the most significantly matching period does not match all of the TOA differences then progressively removing the TOAs with the largest uncertainty will usually increase the significance of the match onto the true period. As a rule of thumb ~8 pulses (i.e. 28 TOA differences) in an observation allow a reliable period estimate, i.e. at the correct harmonic. We note that for a TOA difference to be usable in this method both pulses must not be far-separated so as not to

[6] TOA differences can only be used if the TOAs are from the same observation as we quickly lose phase coherence of TOAs.

Table 4.3 The classifications of the PMPS beams in the PMSingle single-pulse search

Classification	$N_{\text{detections}}$
Candidate: Class 1	162
Candidate: Class 2	204
Candidate: Class 3	319
Known PSR	606 (300)
Known RRAT	13 (11)
Noise	27493
Noise + some remnant RFI	12061
Large remnant RFI	693

For the known sources the number of unique sources detected are given in parentheses

lose coherence, e.g. eight pulses in a single observation yields 28 usable TOA differences for period determination, but two separated observations each of four pulses yields just 12 usable TOA differences. Some example output plots from this method are shown in Fig. 4.12. If a period determination is possible, it can be used with position and DM to further confirm whether or not the candidate is a previously known source. We note one difficulty with this method is that it is possible that the range of emitting pulse longitudes is wide (e.g. for an aligned rotator) so that sharp peaks like those shown in Fig. 4.12 will be smeared out across a wider trial period range.

4.5 Detections and New Discoveries

Currently we have identified 19 sources, 18 of which are new discoveries, the 19th source having also been identified in the PALFA survey [10]. Of these 11, ten of which are new, have been confirmed in numerous followup observations. The other eight sources have not been confirmed despite several re-observation attempts. Nonetheless we consider them to be real astrophysical signals—either low burst rate RRATs or single transient events as described in Sect. 1.3. One of these, J1852−08, is a single burst event which may be extragalactic in origin. All 19 newly identified sources are listed in Table 4.4 with some observed properties. In all sources, except for J1854+03 we know the position to the accuracy of a 1.4-GHz beam of the 64-metre Parkes Telescope, which is 14 arcmin. A more accurate position for J1854+03 is available from Deneva et al. [10]. The DMs given are those obtained from fitting the dispersion sweep of the brightest pulse for each source. The distances quoted are those derived from the DM using the NE2001 model [7, 8] of the electron content of the Galaxy and have typical errors of 20 percent. The pulse widths are those measured at 50% intensity. The quoted 1.4-GHz peak flux densities are determined by using Eq. 4.2 and using the known gain and system temperature of the 20-cm multi-beam receiver (as given in the April 6, 2009 version of the Parkes Radio Telescope Users Guide). The typical uncertainties in this calibration are at the 30% level.

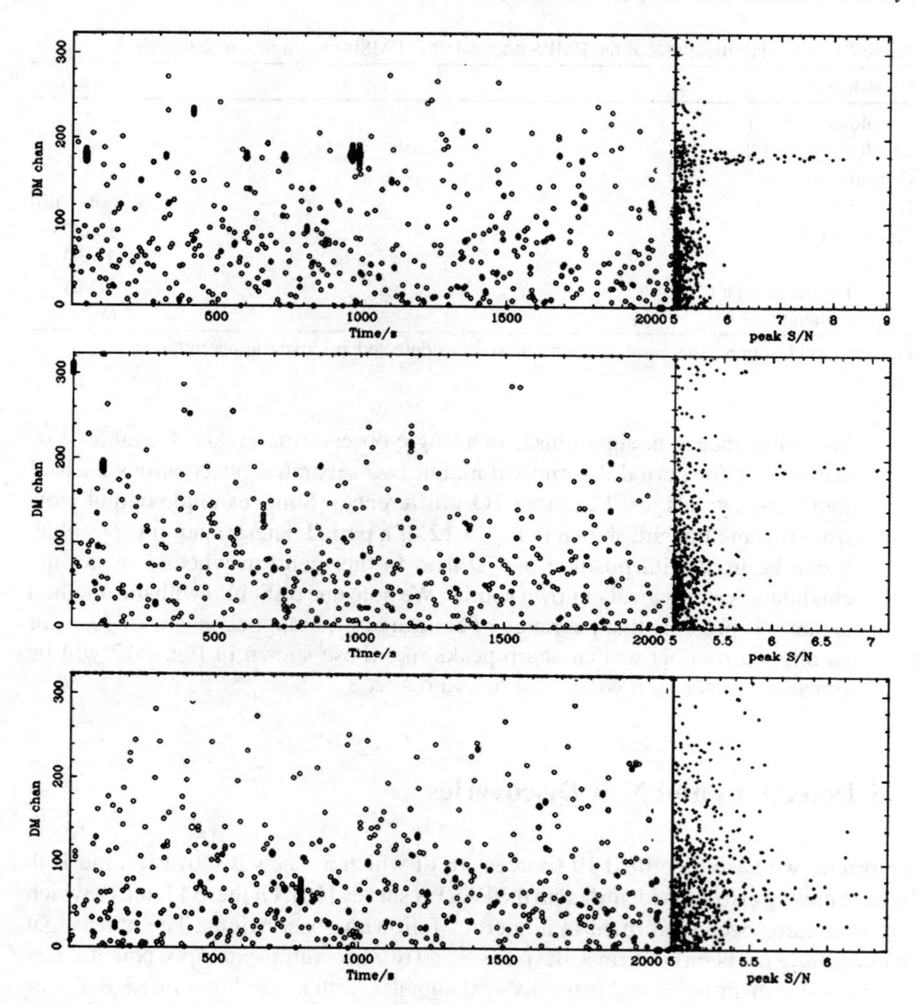

Fig. 4.11 (*Top*) An as yet unconfirmed class 1 candidate with several strong pulses at constant DM which appears to be a real celestial source; (*Middle*) A class 2 source showing one significant pulse; (*Bottom*) A typical Class 3 candidates showing a weak single pulse. Class 3 sources are the least significant with no confirmations expected, especially due to the impractical nature of following up such weak sources

All of these sources were amongst the better class 1 candidates and their detection statistics are given in Table 4.5. Coherent timing solutions have been obtained for all of the repeating sources that have determined periods, but so far spanning only a few months, so that accurate position and period derivative determinations are not yet possible. In this section we describe the detection of previously known sources, the newly discovered sources which have been observed on multiple occasions and the new sources seen in only a single observation.

4.5.1 Detection of Known Sources

In addition to the newly discovered sources, the analysis has made many detections of previously known sources. These include detections of 300 previously known pulsars, often detected multiple times so that there were 606 known pulsar detections in total. Up to 2006, the PMPS detected 976 pulsars (of which, at the time, 742 were new sources, see for e.g. Lorimer et al. [34]). Since then new analyses of results has given rise to more sources such that the ATNF pulsar database now lists 1030 pulsars as detected in the PMPS [11, 23]. The 300 known pulsar detections here then correspond to ~30% of pulsars detected just from single pulse searches. This is an increase on the 250 pulsars detected in the original single pulse search [40] with extra detections made across the entire DM range. The distributions of detected pulsars with respect to DM and period for both analyses are shown in Fig. 4.13.

We can see that extra detections have been made in each DM range. A simple expectation is that the cumulative number of pulsars goes as D^3 until the distance from Earth reaches the Galactic scale height, and for further distances the number goes as D^2. The inverse square law then implies the number of pulsars to grow linearly until a distance of the scale height, after which the number is constant. This expectation is somewhat naive however, not least because n_e in the Galaxy varies hugely along different lines of sight, e.g. scattering and dispersion shroud our abilities to detect sources in the Galactic centre. The peak in the number of pulsars detected corresponds to the Sagittarius spiral arm (see Fig. 4.13), near the DM $= 100\,\mathrm{cm}^{-3}$ pc contour from the NE2001 electron density model, towards the Galactic centre. The Galactic distribution of pulsars is thought to rise further and peak closer to the Galactic centre—at a Galactocentric radius of ~3–5 kpc [34] but the observed distribution falls off at higher DM values, consistent with the expected loss due to increased dispersion and scattering.

The true PMSingle detection rate for known pulsar detections may even be better than 30% if we consider whether any of the pulsars detected in the PMPS would actually be removed by the zero-DM filter. This could be the case for low-DM pulsars. However, if we assume zero-DM must remove the amplitude spectrum up to a fluctuation frequency which is δ^{-1} harmonics in order to remove all information of the pulsar, where $\delta = w/P$ is the pulsar duty cycle, then just 4 of the PMPS pulsars would be removed. Assuming we need to remove just $P/2w$ harmonics would see 16 sources (or 1.6%) removed by the zero-DM filter which leaves the detection rate at ~30%. In addition some of the sources classified as candidates may turn out to be (far side-lobe detections of) known pulsars which could potentially boost the number of single pulse detections by a few percent.

The original 11 RRATs were all re-detected. RRAT J1819–1458 is observed three times in the survey. The third PMPS detection, revealed in this analysis, was previously unknown. This is very helpful for timing and for attempts in connecting over a long ~1800 day gap in timing data (between survey observations and the initial single pulse search of the data). This enables the conclusion that RRAT J1819–1458 does not seem to have suffered a large glitch during this gap in observations, as we

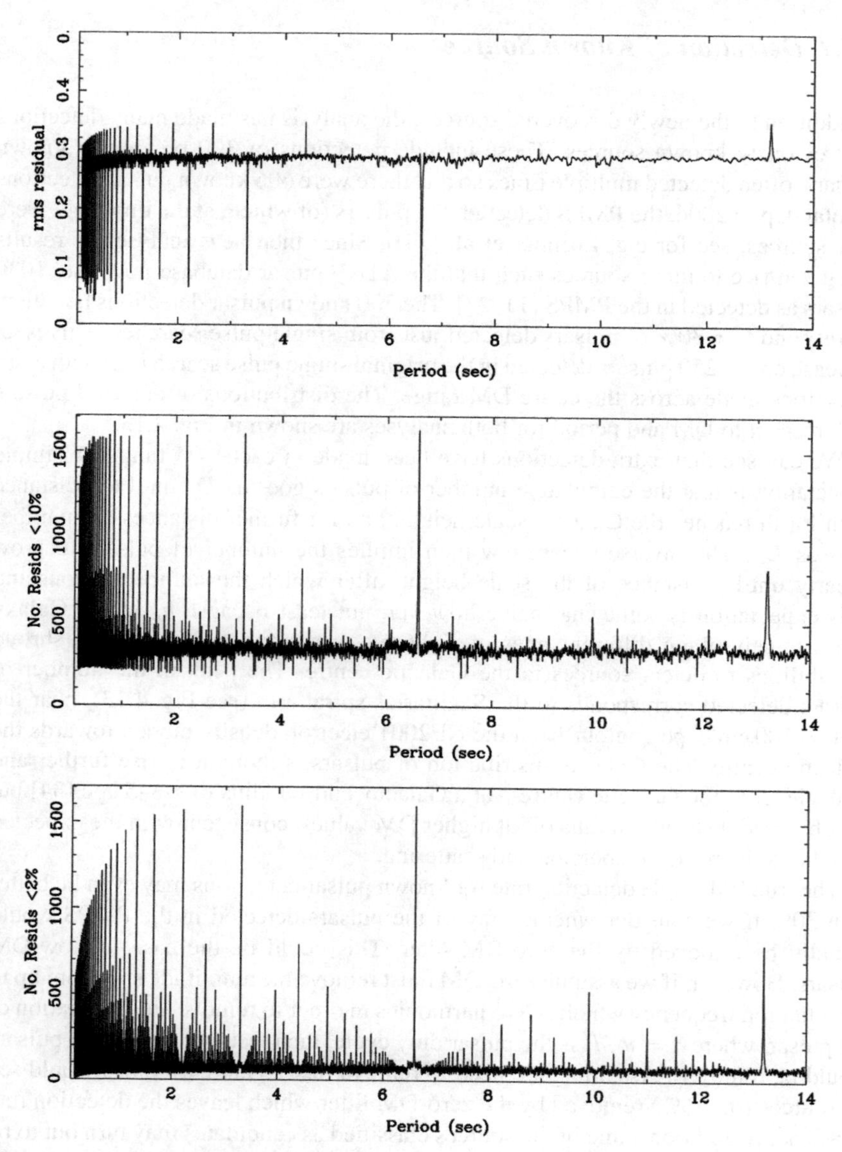

Fig. 4.12 Trial period differences for J1841−14. Here residual is the difference between the expected pulse arrival time (for the assumed trial period) and the actual pulse arrival time. The top panel shows the rms residuals for each trial period. The middle panel shows the number of rms residuals below 10% and the bottom panel shows the number with rms residuals below 2%

will discuss in Chap. 5. A further detection of J1754–30, a year earlier than what was thought to be the first detection, was also identified.

Table 4.4 The observed properties of the newly identified sources from the PMSingle analysis

Source	RA (J2000)	DEC (J2000)	DM (cm^{-3} pc)	D (kpc)	P (s)	Epoch (MJD)	w (ms)	S_{peak} (mJy)	L_{peak} (Jy kpc^2)
Repeating sources									
J1047−58	10:47(1)	−58:41(7)	69.3(3.3)	2.3	1.23129(1)	55779	3.7	630	3.3
J1423−56	14:23(1)	−56:47(7)	32.9(1.1)	1.3	1.42721(7)	54557	4.5	930	1.5
J1514−59	15:14(1)	−59 : 52(7)	171.7(0.9)	3.1	1.046109(4)	54909	3.3	830	7.9
J1554−52	15:54(1)	−52:10(7)	130.8(0.3)	4.5	0.12522947(7)	54977	1.0	1400	28.3
J1703−38	17:03(1)	−38:12(7)	375.0(12.0)	5.7	–	–	9.0	160	5.1
J1707−44	17:07(1)	−44:12(7)	380.0(10.0)	6.7	5.763752(5)	54999	12.1	575	25.8
J1724−35	17:24(1)	−35:49(7)	554.9(9.9)	5.7	1.42199(2)	54776	5.9	180	5.8
J1727−29	17:27(1)	−29:59(7)	92.8(9.4)	1.7	–	–	7.2	160	0.4
J1807−25	18:07(1)	−25:55(7)	385.0(10.0)	7.4	2.76413(5)	54987	4.0	410	22.4
J1841−14	18:41(1)	−14:18(7)	19.4(1.4)	0.8	6.597547(4)	54909	2.6	1700	1.0
J1854+03	18:54:09(7)	+03:04(2)	192.4(5.2)	5.3	4.557818(6)	54909	15.8	540	15.1
Non-repeating sources									
J0845−36	08:45(1)	−36:05(7)	29(2.0)	0.4	–	–	2.0	230	0.04
J1111−55	11:11(1)	−55:19(7)	235(5.0)	5.6	–	–	16.0	80	2.5
J1308−67	13:08(1)	−67:03(7)	44(2.0)	1.2	–	–	2.0	270	0.4
J1311−59	13:11(1)	−59:19(7)	152(5.0)	3.1	–	–	16.0	130	1.3
J1404−58	14:04(1)	−58:15(7)	229(5.0)	4.8	–	–	4.0	220	5.1
J1649−46	16:49(1)	−46:13(7)	394(10.0)	5.1	–	–	16.0	135	3.5
J1652−44	16:52(1)	−44:02(7)	786(10.0)	8.4	7.70718(1)	50835	64.0	40	2.9
J1852−08	18:52(1)	−08:29(7)	745(10.0)	$\sim 5 \times 10^5$	–	–	4.0	410	$\sim 10^{11}$

Table 4.5 Detection statistics for the newly identified sources from the PMSingle analysis. $\dot{\chi}$ refers to the detected burst rate

Source	N_{det}/N_{obs}	N_{pulses}	T_{obs} (hr)	$\dot{\chi}$ (hr^{-1})
Repeating sources				
J1047−58	8/15	54	8.96	6.0
J1423−56	9/12	35	10.01	3.4
J1514−59	9/9	92	4.58	20.0
J1554−52	8/8	214	4.25	50.3
J1703−38	5/6	10	3.08	3.2
J1707−44	5/5	22	2.58	8.5
J1724−35	12/17	34	9.95	3.4
J1727−29	2/5	2	2.11	0.9
J1807−25	7/7	25	3.97	6.2
J1841−14	13/13	231	5.01	46.0
J1854+03	9/9	42	4.52	9.2
Non-repeating sources				
J0845−36	1/2	2	1.09	1.8
J1111−55	1/7	2	4.29	0.4
J1308−67	1/5	2	3.08	0.6
J1311−59	1/6	1	3.29	0.3
J1404−58	1/10	7	6.08	1.1
J1649−46	1/4	1	3.38	0.3
J1652−44†	1/21	9	13.14	0.7
J1852−08	1/8	1	3.70	0.2

† The detection statistics are up to date as of August 2010 for the non-repeating sources. The reasons why there are so many more observations of J1652−44 will be discussed in Sect. 6.1.1.

4.5.2 New Discoveries: Repeating Sources

J1047–58 was found within the same beam as PSR J1048–5832. This allowed for a rapid confirmation of the source by examining archival Parkes data of the known pulsar. In the discovery observations, all of the six detected pulses were clustered within a ~100 second window. In the followup observations of this source there is a suggestion of such 'on' times in which pulses are clustered together in windows of up to ~500 s.

J1423−56 has a rather low dispersion measure, lower than ~85% of known radio pulsars. Detection in the previous analysis was difficult due to the presence of RFI. However this was much improved by the application of our RFI mitigation schemes. From this and followup observations a period of 1.427 seconds has been determined. For this source the detected pulses appear to follow a random distribution, with no evidence for any 'on' windows. We note also that occasionally there is a second component seen 50 ms before the main pulse.

J1514−59 is detected in all observations with a period of 1.046 s and a large average burst rate of $\dot{\chi} \sim 20\,\mathrm{h}^{-1}$. It is not seen in FFT searches of entire observations which have all been ~30 minutes in duration. However its pulses are seen to come in

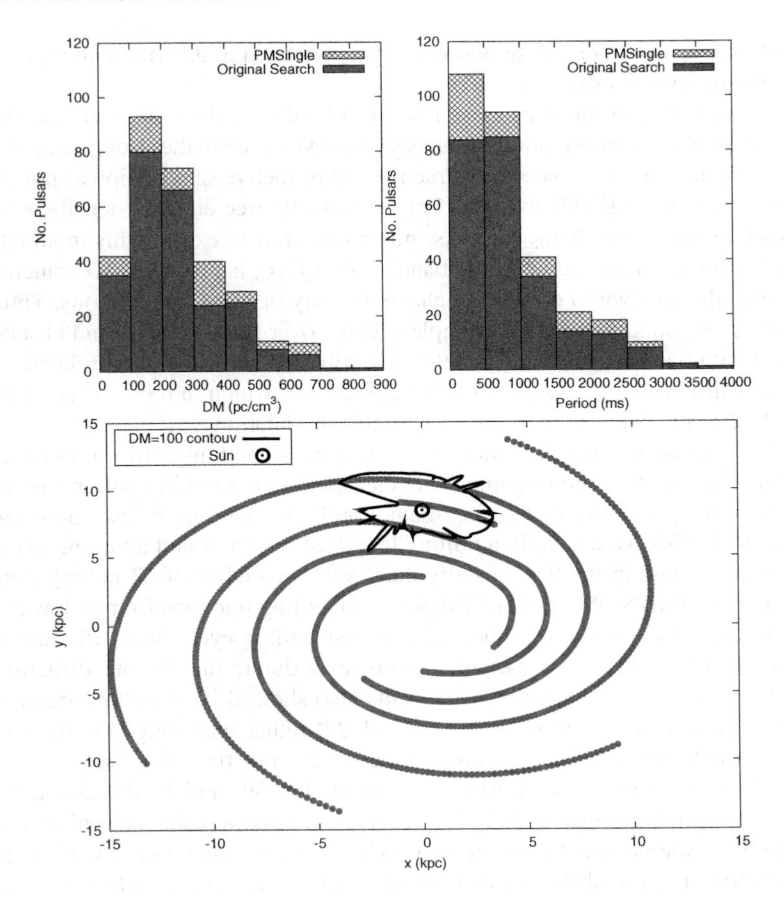

Fig. 4.13 (*Top*) The number distributions of detected known pulsars in this analysis and the original analysis, with respect to DM and period. (*Bottom*) A top-down view of the Milky Way showing the spiral arm structure and the DM = 100 cm^{-3} pc contour of the NE2001 model [7, 8]

approximately minute-long clumps separated by ~800–1000 seconds. Figure 4.10 shows an example of this. Performing FFT searches focused on a small 'on' region (albeit with quite poor spectral resolution of ~10 mHz) gives a period which agrees with that obtained from examining the differences in pulse pair arrival times. Folding the 'on' regions at the nominal period gives us a pulse profile which shows a single narrow-peaked pulse.

Analysing the intervals between the bursts shows that they do not obey a Poisson distribution. The K–S test probability that the distribution is Poissonian is $<10^{-8}$. In fact it seems the distribution is bi-modal with a peak at short burst intervals (of a few periods) and another at long intervals (several hundred periods). We find that the short intervals are consistent with a Poissonian distribution with average expected interval of $\lambda = 7$ periods. More observations are needed to determine the 'peak' of the longer interval distribution.

The long ~15 min intervals do not seem to be due to the effects of interstellar scintillation. For this observation frequency and DM we are in the strong scintillation regime [35] and so must consider diffractive and refractive scintillation as possibilities. Assuming the NE2001 model [7] of the Galactic free electron density we find a diffractive time-scale $\Delta t_{\rm DISS} \sim 30$ s, much too short to explain this modulation. Similarly, the diffractive scintillation bandwidth $\Delta f_{\rm DISS}$ is just ~10 kHz, much narrower than the bandwidth of a single channel in any of these observations. Thus all channels corresponding to a time sample average over many scintles and observing this modulation is not possible. The refractive scintillation timescale is related to the diffractive timescale by $\Delta t_{\rm RISS} = (f/\Delta f_{\rm DISS})\Delta t_{\rm DISS}$ which in this case is ~10 s of days which is much too long to account for this modulation.

It would seem then that this modulation may be something intrinsic to the neutron star. The situation seems somewhat consistent with a nulling pulsar where the majority of the pulses emitted during the non-nulling phase are below our sensitivity threshold. This would imply a nulling length of ~14 min and a nulling cycle of ~1 min, i.e. a nulling fraction of more than 90%. In studies of 23 nulling pulsars detected in the PMPS, Wang et al. [48] observed nulling fractions from as low as 1% to as high as 93%. One source showed a similar nulling cycle of ~515 s although most were lower. This is not inconsistent however due to the obvious difficulty of detecting long-duration nullers. Their results also showed large nulling fraction to be related to large characteristic age, $\tau = P/2\dot{P}$ rather than long periods, a relationship which can be tested for this source once a full timing solution has been obtained. We do note the similarity between J1514−59 (and J1047−58) and the class of "intermittent pulsars" [29]. However the time-scales for the 'on' and 'off' states in these sources can be 10 s of days which is much longer than what is seen in these RRATs. The possibility remains though that RRATs may fit into a continuum of nulling behaviour which could range from those sources which null for a few periods at a time at one extreme to the intermittent pulsars at the other extreme. As the numbers of known RRATs and intermittent pulsars increases timing observations can be used to investigate what properties (if any, e.g. period and age) correlate with nulling fraction.

J1554−52 is a strong single pulse emitter showing 35 pulses in its discovery observation. It is also weakly detectable in FFT searches of most observations but with much higher significance in single pulse searches. The weakness in the FFT detection is one reason for the previous non-detection of this source. However it is likely that this source would have been removed by previously applied algorithms designed to remove RFI signals. For instance frequency domain 'zapping' would have removed this source, i.e. setting certain frequencies in the fluctuation spectrum (of a dedispersed time series) to zero. This is done at known RFI frequencies (e.g. 50 Hz) and their harmonics. With a period of 125 ms this pulsar falls exactly into one of these zapped regions.

J1703−38 is observed once in the PMPS and has been re-observed in all but one of the five followup observations performed so far. With only ten pulses detectable in total a period determination has not been possible. Thus there remains the possibility that this source is a known pulsar detected in a side-lobe of the telescope beam on the sky, although where multiple pulses are detected in an individual observation the pulse time-of-arrival differences are not consistent with the periods of known pulsars in this region of the sky. More observations will yield a period for this source (see Sect. 6.1.2).

J1707−44 is observed once in the PMPS and in four followup observations. The period of this source is long compared to most radio pulsars at 5.764 s. No evidence is seen for any 'clumping' of pulses as in J1514−59 and pulses all seem to come from a single phase window.

J1724−35 was the first source to be discovered in this re-processing [12]. Since the two survey observations of this source it has been re-observed 15 times and detected in ten of those. Despite having a fairly high DM its discovery is helped immensely by the removal of very strong RFI by the zero-DM filter. In all but two observations it is not detected in an FFT search. In one observation it can be detected from focused FFT searches of times when strong single pulses are seen, as for J1514−59. In another observation it is detected with FFT S/N of 15 which is evidence that there is underlying weak emission in addition to the detected single pulses. During these times when the source is detectable in periodicity searches a folded profile can be obtained which is quite wide and double-peaked. We note that this variation is not due to scintillation as the scintillation time-scale of two seconds is too short and the bandwidth of ~20 Hz is too narrow to explain this. The burst rate is insufficient for analysis of the intervals between bursts at present.

J1727−29 is detected once in the PMPS with just one strong single pulse. It was re-observed in a followup observation where we again detected just one pulse. Further followup observations have not revealed anything further from this source. This source is obviously too weak to time and without even two pulses in a single observation a period estimate is not possible. We expect a number of candidates like this to be confirmed while proving impractical to continuously monitor. Such sources will require the next-generation sensitivity of the SKA and further serious followup efforts should be engaged when such facilities become available.

J1807−25 is observed once in the PMPS and in all of the six subsequent followup observations at Parkes. It has a rotation period of 2.764 s and, so far, seems to have only one pulse component detected. It's pulse rate should be sufficiently high to allow the determination of a timing solution.

J1841−14 was observed twice in the PMPS, where, as we can see from Fig. 4.8, its detection was hindered by the presence of strong RFI in the initial analysis. The source has been re-observed 11 times and detected in all cases. It has the lowest DM of any RRAT found in the PMPS (and lower than 95% of normal radio pulsars). It has a very high burst rate of $\dot{\chi} \sim 40\,\mathrm{h}^{-1}$ but it is undetectable in an FFT search. However this seems to be due to the insensitivity of FFT searches in detecting long periods.[7] Using a Fast Folding Algorithm [25], an alternative periodicity search, more sensitive

[7] This is due to red-noise and in the case of the zero-DM filter a suppressed fluctuation spectrum at low frequencies.

than FFTs for high periods, we detect the source at the correct period. The average FFA S/N is 9 as compared to the peak single pulse S/N of ~60 and many pulses per observation with single pulse S/N ≥15. It has a very long period of 6.596 s as determined from factorising the pulse pair time-of-arrival differences. Figure 4.12 shows the results of this period determination. Folding the observations at this period shows a narrow pulse profile.The pulses observed are among the brightest seen for RRATs with typical peak flux densities (at 1.4 GHz) of ~1 Jy and a maximum peak flux density observed of 1.7 Jy. While most of the pulses are narrow at ~2 ms there are few pulses detected with pulse widths of as wide as 20 ms. The high burst rate means that obtaining a sufficiently large number of TOAs at regular intervals to obtain a coherent timing solution should be straightforward. Obtaining an accurate timing position will be useful for a detection of this source at X-rays which appears very promising as the source is nearby at a distance of ~800 pc, a possibility which is discussed further in Chap. 6.

J1854+03 was observed once in the PMPSand has since been re-observed eight times and detected in allcases. This source is one previously identified by the 1.4-GHz PALFA survey [10]. As the PALFA position is much more accurate than that which we were able to obtain with the Parkes telescope (due to the much smaller beam size of the Arecibo telescope) it may be possible to determine a period derivative for this source on a shorter timescale than for the other sources. This is because, typically, determining a period derivative takes a year of timing observations so that the effects of positional uncertainty (which shows year-long sinusoidal patterns in timing residuals) and the slow-down rate of the star can be disentangled. This source has a high burst rate which is $\dot{\chi} \sim 10\ \mathrm{h}^{-1}$. It is undetectable in FFT searches and has a long period of 4.558 seconds. However this more distant source (it is ~6 times further away than J1841−14) shows weak pulses. Typical peak flux densities (at 1.4 GHz) are ~100 mJy but the brightest observed pulse is ~540 mJy. The pulse widths are typically ~15 ms and there are no indications of clumps of emission on which to focus FFT searches.

4.5.3 New Discoveries: Non-Repeating Sources

In addition to the confirmed sources described here we have identified a number of candidates which we deem to be 'self-confirmed', i.e. we have not re-observed bursts in a followup observation but we deem the survey detection sufficiently convincing that the astrophysical nature of these sources is clear. A number of these are just single bursts, show the characteristic dispersive delay expected from celestial sources, are detected in only one beam and show no signatures of RFI. Figures 4.14 and 4.15 show DM–time search output and frequency-time plots showing the dispersive sweep of individual pulses, for some of these sources. We can see that these sources obey the theoretical dispersion law of Eq. 4.23.

The sources show between 1 and 9 pulses in their discovery observations and have been followed up for between ~1 and ~6 h, without showing further pulses.

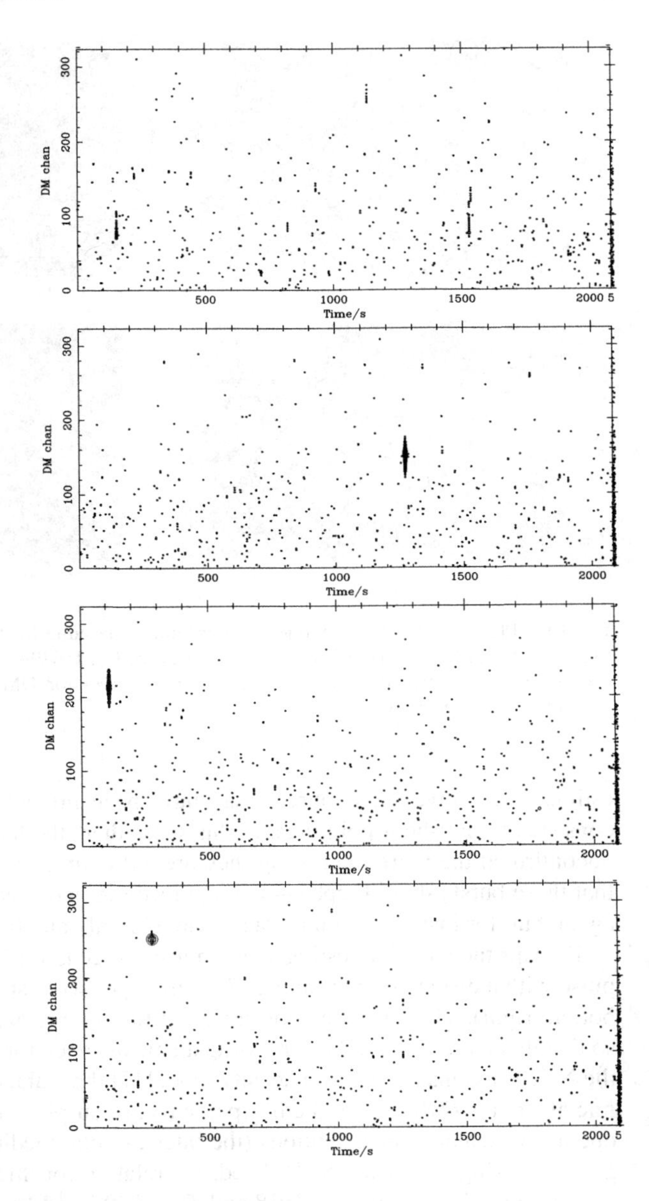

Fig. 4.14 DM–time search plot output for four of the single observation PMPS RRATs. These are, from *top* to *bottom*, J1308−67, J1311−59, J1649−46 and J1852−08

As these bursts are just a few milliseconds in duration a neutron star or, perhaps, a more compact object is expected to be the source of the emission. As the discovery observations clearly show these sources to be astrophysical, the long followups with no confirmation suggest a very slow rate of bursting. These sources are then of huge relevance when considering the population size of such sources. For instance, if a source shows one burst in 5 h of observation, it suggests that, as a zeroth order

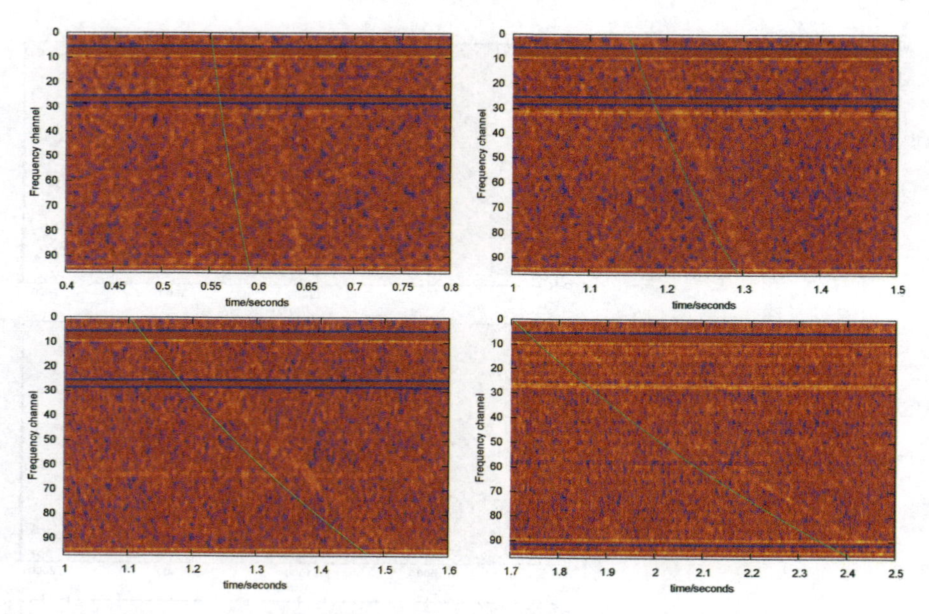

Fig. 4.15 Plotted is the time–frequency–amplitude data cube for the single pulses identified in Fig. 4.14. for J1308−67 (*top left*), J1311−59 (*top right*), J1649−46 (*bottom left*) and J1852−08 (*bottom right*). Overplotted are theoretical curves *green* for DM values of 44, 152, 394 and 745 cm^{-3} pc respectively

estimate, that nine such sources may have been missed during the survey which consisted of half-hour pointings. In this sense then, the longer these sources remain unconfirmed the more interesting they are. Of course we have no reason to expect that these bursts should repeat—a temporarily re-activated 'dead' pulsar may burst again, but, for instance, a mini black hole will only annihilate once!

Perhaps the most interesting single pulse source is J1852−08, an isolated 4.0-ms pulse with a dispersion measure of 745 cm^{-3} pc (see dispersive sweep in Fig. 4.15, bottom right). The Galactic coordinates of this source are $l = 25.4°$, $b = -4.0°$, so that this large DM implies an extragalactic distance for this source. According to the NE2001 Galactic electron density model [7] the Galactic contribution along this line of sight is $DM_{\mathrm{Gal}} = 533\ \mathrm{cm}^{-3}$ pc, leaving a surplus of $DM_{\mathrm{extra}} = 222\ \mathrm{cm}^{-3}$ pc due to extragalactic contributions (the inter-galactic medium and any putative host galaxy). Using the Ioka [18] DM-redshift relation for this component, the inferred redshift and distance are $z \approx 0.18$ and $D \approx 520\ \mathrm{h}^{-1}$ Mpc ($\Omega_{\mathrm{m}} = 0.3$, $\Omega_{\Lambda} = 0.7$), if all of the DM_{extra} component is due to the inter-galactic medium. Allowing for a contribution of 100 cm^{-3} pc from a host galaxy (see Lorimer et al. [33]), these values become $z \approx 0.09$ and $D \approx 260\ \mathrm{h}^{-1}$ Mpc. Of course there may be some unknown contribution to the free electron density on the far side of the Galactic centre which, if taken into account, might reduce the inferred distance to within the Galaxy. Either way, J1852−08 has the strongest intrinsic peak luminosity of all the PMPS sources:

Table 4.6 Observed characteristics of (non-recyled) pulsars, magnetars and RRATs

Property	Pulsar ($\sim$1700 sources)	Magnetar (18 sources)	RRAT (29 sources)
Period (s) [range, median]	$\sim$0.03–8.5, 0.54	2.1–11.8, 7.3	0.1 – 6.7, 2.1
log (Period derivative)	-17 to -12	-13 to -9 (14 measured)	-15 to -13 (7 measured)
$\log(\dot{E})$ (erg/s)	30–38	31–35	31–33
$\log(B_S)$ (gauss) [range, median]	10.0–14.0, 12.03	13.0–15.0, 14.5 $> B_{\text{Quantum}}$	12.4–13.7, 13.0
$\log(B_{LC})$ (gauss)	-2 to 6	~ -0.01 to 1.1	~ -0.1 to 1.7
$\log(\tau = \dot{P}/2P)$ (yr)	3–9	2–6	5–6
Observed radio on-fraction	0.05 – 1	0–1	$10^{-3} - 10^{-2}$
Transient behaviour	Nulls, intermittent PSRs	X-ray and radio variability	$\dot{\chi} \sim 1\ \text{min}^{-1}$ to $\sim 1\ \text{hr}^{-1}$
Bursting behaviour	GRPs, giant micro-pulses, nano-giant pulses	X-ray and soft γ-ray bursts	$\lesssim$10 Jy single pulses
No. of components/subpulses	$\geq$1 components	Large phase range, many components in XTE 1810$-$197	$\geq$1 components $\geq$1
Beaming fraction ($f_{\text{beam}}(P)$)	$0.09[\log(P/s) - 1]^2 + 0.03$	Unknown	Unknown
Amplitude distribution	GRP: Gaussian + power law, indices -1.5 to -4	Power-law or lognormal, depending on which component	Some lognormal (see Sect. 4.6.2), some powerlaw, index -1
Radio spectra	$\nu^{-\alpha}$, $\langle\alpha\rangle = 1.8 \pm 0.2$	Flat but variable	Unknown
X-ray spectra	Thermal and/or power-law	Thermal + power law	Thermal in J1819$-$1458
Polarisation	Low–high	$\lesssim$100 percent	$L/I \sim 0.37$, $V/I \sim 0.06$ for J1819$-$1458

Information in this table is taken from many sources: in addition to those references mentioned in the main text these include Camilo et al. [3–5]; Karastergion et al. [20]; Lazaridis et al. [32]; Maron et al. [39]; McLaughlin et al. [42]; Serylak et al. [46] and the McGill Magnetar Catalogue http://www.physics.mcgill.ca/~pulsar/magnetar/main.html. B_{LC} and B_S refer to the inferred magnetic fields at the light cylinder and stellar surface respectively assuming a star with a dipolar magnetic field. B_{quantum} is the quantum-critical magnetic field above which the separation of Landau levels for synchrotron orbits exceeds the electron rest mass. The quantities L, V and I refer to linear polarisation, circular polarisation and total intensity respectively. We note that although there are 29 published RRAT sources there are determined periods for just 27 and spin-down properties like $\dot{P}$, B_S, $\dot{E}$ and τ are known for just seven sources [42]

Table 4.7 The required distances to see continuous emission if these sources emit according to various power laws

Source	g	D (kpc)	$g = 1$ distance (kpc)			
			($\alpha = 1.5$)	($\alpha = 2$)	($\alpha = 3$)	($\alpha = 4$)
J1047−58	0.0021(1/476)	2.3	0.01	0.12	0.50	0.83
J1423−56	0.0014(1/714)	1.3	<0.1	0.05	0.25	0.43
J1514−59	0.0058(1/172)	3.1	0.03	0.26	0.86	1.31
J1554−52	0.0016(1/625)	4.5	0.01	0.20	0.91	1.54
J1707−44	0.0137(1/73)	6.7	0.14	0.84	2.30	3.28
J1724−35	0.0013(1/769)	5.7	0.03	0.30	1.18	1.94
J1807−25	0.0048(1/208)	7.4	0.08	0.61	2.00	3.05
J1841−14	0.0845(1/11.8)	0.8	0.10	0.24	0.43	0.53
J1854+03	0.0118(1/85)	5.3	0.10	0.61	1.75	2.53

$\gtrsim 10^{11}$ Jy kpc^2 if extragalactic (as in Fig. 4.2), or $\gtrsim 10^3$ Jy kpc^2 (20 times stronger than J1819−1458) if at the edge of the Galaxy. Of the sources that showed repeated bursts, J1404−58 showed seven pulses but we are unable to confidently identify any underlying period. J1652−44 shows nine pulses which has allowed us to determine an underlying periodicity of 7.7 s, and this source is discussed in more detail in Sect. 6.1.1.

4.6 Discussion

The motivation for this re-processing of the PMPS was to find more RRAT sources, encouraged by the discovery potential of their large projected population. As discussed in Chap. 3 it is important to truly understand how RRATs fit into our picture of Galactic neutron stars: it may even tell us something about neutron star evolution. It is therefore important to clearly describe differences between RRATs and pulsars. Can we propose a more meaningful definition of what a RRAT is? The answer to this should also elucidate matters when considering the overarching question: Are RRATs special?

4.6.1 What is a RRAT?

The initial de facto definition of what a RRAT is, is the following: A RRAT is a repeating radio source which is detectable via its single pulses and is not detectable in periodicity searches.

This definition is flawed in numerous ways: (1) It is incorrect as we cannot have $r \to \infty$. We might re-state the definition more properly so as to evade this problem:

A RRAT is a repeating source which is more easily detectable via its single pulses as compared to a periodicity search; (2) It depends on the observing time! From Sect. 4.2 we know that $r = r(N) = r(T_{\rm obs}/P)$ with a diverse range of behaviour, as a function of observing time, depending on the amplitude distribution of the source. This means that a 'RRAT' in the PMPS may be detected as a 'normal pulsar' in a different survey with a different observing time, or vice-versa! (3) If a RRAT has $r > 1$ then, even for the same observing time and identical observing setup, we cannot clearly classify a source as a RRAT as r varies from observation to observation. Some kind of arbitrary average r would need to be considered which seems contrived and unhelpful. (4) As we know nothing about the radio spectra of RRATs we face the possibility of a source being identified as a RRAT at one frequency but not at others. Moreover, such definitions are *detection based*, and say nothing definitive about the intrinsic nature of the source. This becomes a problem when trying to make statements about the relationships between each manifestation of neutron star. Below we make an attempt to determine a more meaningful definition incorporating information about intermittency, amplitude and spectral distributions, multi-frequency behaviour as well as period and period derivative properties and the derived quantities related to these. This information is compiled in Table 4.6. *Ultimately the endeavour to formulate an all-encompassing intrinsic definition for RRATs fails*, and, as we discuss in Chap. 8, more than one explanation for the RRAT phenomenon is needed.

4.6.2 Distant Pulsars?

Can we explain RRATs as distant analogues of pulsars having pulse amplitude distribution with a long, high flux density tail? Weltevrede et al. [49, 50] have shown that PSR B0656+14 (at a distance of 288 pc) would appear RRAT-like if moved to typical RRAT distances. The amplitude distribution of the pulses from this source is multi-modal, lognormal but with a power law at the high-flux density end, with α between -2 and -3. We can test this scenario if we assume that RRATs emit pulses according to a power law amplitude probability distribution of the form of Eq. 4.12, between S_1 and S_2. We detect all pulses above $S_{\rm thresh}$ where $S_1 \leq S_{\rm thresh} \leq S_2$. This means that the fraction of observable pulses, g, is given by

$$g = \frac{\int_{S_{\rm thresh}}^{S_2} f(S)dS}{\int_{S_1}^{S_2} f(S)dS} \tag{4.28}$$

where S_2 (which we take to be the strongest observed pulse), $S_{\rm thresh}$ and g are all known. The observed values for g for the 9 repeating PMSingle sources with known periods[8] are given in Table 4.7. Thus for various chosen power law indices α, we can determine S_1. This can be used to determine the distance where we would need

[8] To determine g the period must be known as g is the average number of pulses per period, or $g = (\dot{\chi}/{\rm hr}^{-1})((P/s) \times 3600)$.

to move the RRAT to see all of its pulses (i.e. observe it like a pulsar which emits continuously with $g = 1$) from $D_{\rm new} = D(S_1/S_{\rm thresh})^{1/2}$.

We can see that for the steepest power laws the source is not required to move very much nearer for all pulses to become visible. As the power law index gets shallower the source must be brought ever closer to be seen as a continuous emitter. For $\alpha \lesssim 1.5$, the change in distance becomes unreasonable such that if all RRATs were continuously emitting pulses with energy distributions as a power law with $\alpha \lesssim 1.5$ almost none of these sources would ever appear as continuous emitters for a reasonable distance distribution and such sources would be seen as distinct from pulsars. We thus conclude that RRAT emission could be explained as coming from distant pulsars, i.e. continuous emitters, with steep power-law distributions only. For shallower pulse distributions a power-law alone cannot explain the observed RRAT emission as being due to distant pulsars. However the sources may still be seen as continuous if the distribution were to break to, e.g., a log-normal distribution at low flux densities. We can compare these results with the amplitude distributions from the initial RRAT discovery paper which showed some distributions being consistent with power law indices of $\alpha = 1$ [43]. Clearly, the amplitude distribution of pulses will provide a powerful discriminator between sources that can be explained as distant pulsars and those which cannot.

Figure 4.16 shows the amplitude distributions for J1514−59, J1554−52 and J1841−14, the three sources discussed here with the highest number of detected pulses. These distributions are found not to be consistent with a power-law distribution but instead are well fitted by log-normal distributions, the parameters of which are given in Table 4.8. The best-fit curves are over-plotted on the observed distributions in Fig. 4.16. For these three sources there is a low flux density turn-over. It is not clear whether this is an intrinsic turn-over or simply due to the sensitivity threshold. The flux density threshold for a single pulse depends on the pulse width. Plugging in the known observing parameters for Parkes into the radiometer equation gives a single pulse peak flux of: $S_{\rm peak} \approx 245\,{\rm mJy}(w/{\rm ms})^{-\frac{1}{2}}$ assuming a 5-sigma detection threshold. Although the widths of the pulses vary from pulse to pulse we can take the average widths from Table 4.4 to get sensitivity estimates of 135 mJy, 250 mJy and 150 mJy for J1514−59, J1554−52 and J1841−14 respectively. If the turn-overs were intrinsic to the sources then it would suggest that we are not just seeing the brightest pulses from a continuously emitting source but rather that we are seeing most pulses which are emitted. If this is the case, the bursty behaviour is indeed due to the lack of continuous emission, and is an innate property of the sources. For the remaining sources the number of pulses detected is as yet still too low for such an analysis. Continued observations will allow accurate determination of amplitude distributions for all the sources.

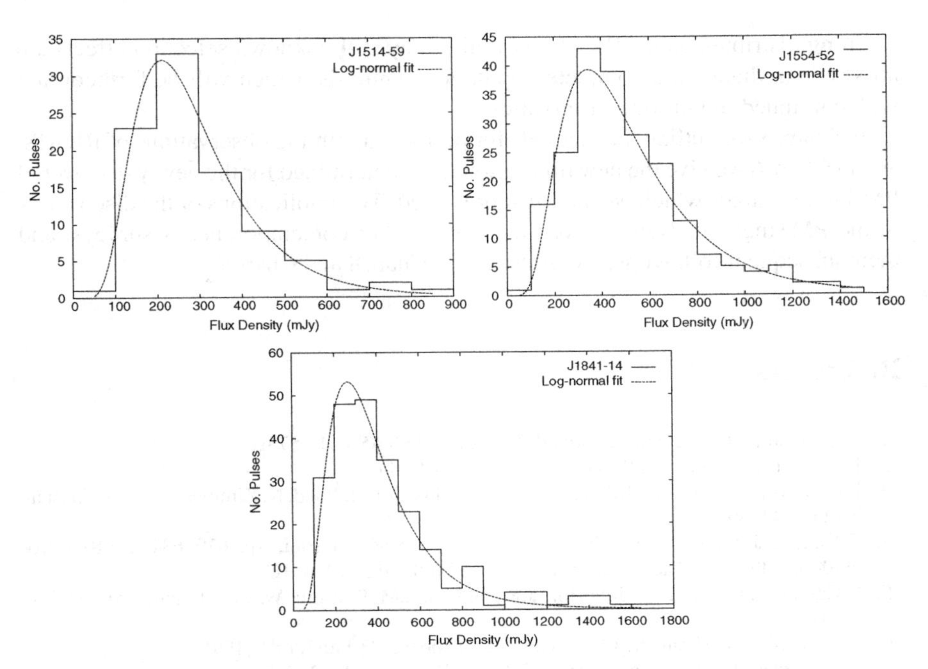

Fig. 4.16 Amplitude distributions for: (*Top*) J1514−59; (*Middle*) J1554−52 and; (*Bottom*) J1841−14

Table 4.8 The best-fit parameters to the amplitude distributions in Fig. 4.16 for a lognormal probability density distribution of the form: $P(x) = (a/x)\exp\left[-\frac{(\ln x - b)^2}{2c^2}\right]$. The parameter a is an arbitrary scaling factor and the values given here correspond to the scales used in Fig. 4.16.

Source	a	b	c
J1514−14	7613(748)	5.57(0.03)	0.47(0.03)
J1554−52	15662(1182)	6.13(0.03)	0.53(0.03)
J1841−14	16130(1704)	5.86(0.04)	0.53(0.04)

4.7 Conclusions

Considering the pointings with strong residual RFI and the fraction of sources removed by zero-DM in the light of the PMSingle analysis the number of beams still affected by RFI is $(41561-693) * 0.016 + 693$ which is ∼3% of all beams, so that 97% are now cleaned of RFI. We have almost tripled the number of known PMPS RRATs (11 → 30) while simultaneously reducing the fraction missed by RFI to almost zero. The inferred population estimate is thus related to the initial estimate by a factor of $(30/11)/(97/50) \approx 1.5$ so that the confirmed number of new sources is consistent with the original population estimate [43]. In fact, the estimates may need to be revised upwards. Although, the as yet undetermined burst rate and beaming distributions for RRATs as well as any still unknown selection effects are all key ingredients in a complete population synthesis, which will be forthcoming with continued monitoring observations.

In Chap. 5 we outline the methods for performing timing observations of RRATs, and in Chap. 6 we give the new timing solutions determined for the newly discovered PMSingle sources which we have just presented. The implications of the discoveries in the PMSingle analysis, as well as several other contemporaneous surveys, and their subsequent followups, are discussed in Chap. 8 and Chap. 9.

References

1. Bates et al., http://adsabs.harvard.edu/abs/2011MNRAS.411.1575B
2. I.H. Cairns, S. Johnston, P. Das, ApJ **563**, L65 (2001)
3. F. Camilo, S.M. Ransom, J.P. Halpern, J. Reynolds, D.J. Helfand, N. Zimmerman, J. Sarkissian, Nature **442**, 892 (2006)
4. F. Camilo, J. Reynolds, S. Johnston, J.P. Halpern, S.M. Ransom, ApJ **679**, 681 (2008). astro-ph/0802.0494, http://adsabs.harvard.edu/abs/2008ApJ...679..681C
5. F. Camilo, J. Reynolds, S. Johnston, J.P. Halpern, S.M. Ransom, W. van Straten, ApJ **659**, L37 (2007)
6. I. Cognard, J.A. Shrauner, J.H. Taylor, S.E. Thorsett, ApJ **457**, L81 (1996)
7. J.M. Cordes, T.J.W. Lazio, preprint (2002). arXiv:astro-ph/0207156
8. J.M. Cordes, T.J.W. Lazio, preprint (2003). arXiv:astro-ph/0301598
9. J.M. Cordes, M.A. McLaughlin, ApJ **596**, 1142 (2003)
10. J.S. Deneva et al., ApJ **703**, 2259 (2009)
11. R.P. Eatough, Ph.D. thesis (University of Manchester, 2009)
12. R.P. Eatough, E.F. Keane, A.G. Lyne, MNRAS **395**, 410 (2009)
13. R.P. Eatough, N. Molkenthin, M. Kramer, A. Noutsos, M.J. Keith, B.W. Stappers, A.G. Lyne, MNRAS (2010) (in press)
14. A.J. Faulkner et al., MNRAS **355**, 147 (2004)
15. T.H. Hankins, J.A. Eilek, ApJ **670**, 693 (2007)
16. J. Hessels et al., ArXiv e-prints (2010). astro-ph/1009.1758
17. G. Hobbs et al., MNRAS **352**, 1439 (2004)
18. K. Ioka, ApJ **598**, L79 (2003)
19. S. Johnston, W. van Straten, M. Kramer, M. Bailes, ApJ **549**, L101 (2001)
20. A. Karastergiou, A.W. Hotan, W. van Straten, M.A. McLaughlin, S.M. Ord, MNRAS **396**, 95 (2009) (astro-ph/0905.1250)
21. R. Karuppusamy, B.W. Stappers, W. van Straten, **515**, A36
22. Keane et al., MNRAS **401**, 1057–1068 (2010)
23. M.J. Keith, R.P. Eatough, A.G. Lyne, M. Kramer, A. Possenti, F. Camilo, R.N. Manchester, MNRAS **395**, 837 (2009)
24. P. Weltevrede, K.P. Watters, B.W. Stappers, MNRAS **389**, 1881 (2008)
25. V.I. Kondratiev, M.A. McLaughlin, D.R. Lorimer, M. Burgay, A. Possenti, R. Turolla, S.B. Popov, S. Zane, ApJ **702**, 692 (2009)
26. M. Kramer, Ph.D. thesis (University of Bonn, 1995)
27. M. Kramer et al., MNRAS **342**, 1299 (2003)
28. M. Kramer, A. Jessner, P. Müller, R. Wielebinski, in *Pulsars: Problems and Progress*, ed. by S. Johnston, M.A. Walker, M. Bailes. IAU Colloquium 160, vol. 1 (Astronomical Society of the Pacific, San Francisco, 1996), p. 13
29. M. Kramer, A.G. Lyne, J.T. O'Brien, C.A. Jordan, D.R. Lorimer, Science **312**, 549 (2006)

30. M. Kramer, B. Stappers, in ISKAF2010 Science Meeting (2010). astro-ph/1009.1938
31. M. Kramer, K.M. Xilouris, A. Jessner, M. Wielebinski, R. Timofeev, A&A **306**, 867 (1996)
32. K. Lazaridis, A. Jessner, M. Kramer, B.W. Stappers, A.G. Lyne, C.A. Jordan, M. Serylak, J.A. Zensus, MNRAS **390**, 839 (2008)
33. D.R. Lorimer, M. Bailes, M.A. McLaughlin, D.J. Narkevic, F. Crawford, Science **318**, 777 (2007)
34. D.R. Lorimer et al., MNRAS **372**, 777 (2006)
35. D.R. Lorimer, M. Kramer, Handbook of Pulsar Astronomy. Cambridge University Press
36. S.C Lundgren, J.M. Cordes, M. Ulmer, S.M. Matz, S. Lomatch, R.S. Foster, T. Hankins, ApJ **453**, 433 (1995)
37. V.M. Malofeev, O.I. Malov, N.V. Shchegoleva, Astron. Rep. **44**, 436 (2000)
38. R.N. Manchester et al., MNRAS **328**, 17 (2001)
39. O. Maron, J. Kijak, M. Kramer, R. Wielebinski, A&AS **147**, 195 (2000)
40. M. McLaughlin, (2009) in *Astrophysics and Space Science Library*, ed. by W. Becker, ASSL, vol. 357, p. 41
41. M.A. McLaughlin, J.M. Cordes, ApJ **596**, 982 (2003)
42. M.A. McLaughlin et al., MNRAS **400**, 1431 (2009)
43. M.A. McLaughlin et al., Nature **439**, 817 (2006)
44. D.J. Morris et al., MNRAS **335**, 275 (2002)
45. J.T. O'Brien et al., MNRAS **388**, L1 (2008)
46. M. Serylak et al., MNRAS **394**, 295 (2009)
47. J. van Leeuwen, B.W. Stappers, A&A **509**, A7 (2010)
48. N. Wang, R.N. Manchester, S. Johnston, MNRAS **377**, 1383 (2007)
49. P. Weltevrede, B.W. Stappers, J.M. Rankin, G.A.E. Wright, ApJ **645**, L149 (2006)
50. P. Weltevrede, G.A.E. Wright, B.W. Stappers, J.M. Rankin, A&A **458**, 269 (2006)

Chapter 5
Timing Observations of RRATs

In the following chapter, parts of Sect. 5.2.2 have been published in the Monthly Notices of the Royal Astronomical Society [17]. Sections 5.1.1 and 5.2.2 constitute parts of two papers, currently in preparation. The chapter sets out the methods used in timing RRATs.

By the beginning of 2008, the original 11 RRATs had been monitored regularly for $\sim$3–4 years at Parkes and there were three sources with coherent timing solutions. It turned out that the originally published period for J1754$-$30 was in error by a factor of 3 ($P = 1.26$ s as opposed to 0.42 s) and the timing solution for J1913+13 was actually slightly incorrect (as discussed below in Sect. 5.2.3). In this chapter we describe the methods involved in, and the difficulties encountered when, timing RRATs. We discuss the timing behaviour of J1819$-$1458 in particular.

5.1 Pulsar Timing

Pulsars are commonly referred to as stable astrophysical clocks. However, even though they are rotationally stable, on a period-by-period basis the pulses we detect from pulsars are quite variable in amplitude, phase and shape. These individual pulses (aka sub-pulses) can vary in random as well as highly ordered ways. Sub-pulse drifting is a phenomenon whereby the rotational phase wherein we see pulsar emission changes periodically (see e.g. Weltevrede et al. [27]). Some pulsars also exhibit 'mode-changing', or 'moding', whereby they are seen to switch between two or more different stable emission profiles [5]. Another phenomenon is nulling, which can be seen as an extreme example of moding, where one of the modes shows no radio emission, i.e. the radio emission ceases and the pulsar is 'off' [26]. Random changes are usually labelled as 'pulse jitter', e.g. the Gaussian variations in pulse phase seen in PSR J0437$-$4715 (K. Liu et al., in preparation). These phenomena are discussed in more detail in Chap. 8. For the purposes of 'timing' a pulsar, i.e. modelling its rotational phase as a function of time with respect to pulsar and astrometric

E. F. Keane, *The Transient Radio Sky*, Springer Theses,
DOI: 10.1007/978-3-642-19627-0_5, © Springer-Verlag Berlin Heidelberg 2011

parameters, these variations all amount to 'timing instabilities'. We note that none of the effects mentioned above are symptomatic of *rotational* irregularities—the pulsar is still spinning down in a well-behaved manner, what is variable/unstable is the source of the radio emission. There are also rotational instabilities known as glitches which are single events consisting of instantaneous jumps in rotation frequency and its derivatives. These are further described in Sects. 5.2.1 and 5.2.2. The, possibly more general, phenomena of slow-down rate switching may be occurring in much of the pulsar population [16].

5.1.1 Integrated Profiles

To perform 'pulsar timing' of a source it is usually observed for a large number of pulse periods. The observation is integrated to create an average pulse profile $P(t)$. The addition of many pulse periods is performed for two reasons: (1) To compensate for all of the timing instabilities outlined above, and (2) To increase the signal-to-noise ratio of $P(t)$. It is sometimes mistakenly assumed that a high signal-to-noise ratio implies a stable profile but this is not true (we define stability below). In practise, as many periods as possible are used in timing 'normal'/'slow' pulsars, typically 10^2–10^3, but for the faster millisecond pulsars (MSPs) $\gtrsim 10^5$ are used routinely. Determining a pulse time-of-arrival (TOA) for a given observation then amounts to cross-correlating the observed profile $P(t)$ with a very high S/N (or sometimes even analytic) template profile $T(t)$ under the *assumption* that the profile is just a shifted, scaled and noisier version of the template, i.e.

$$P(t) = AT(t + \psi) + N(t), \tag{5.1}$$

where A is a scale factor, ψ is a phase shift and N is an additive noise term. Determining ψ gives the TOA relative to some known reference time, usually the observatory clock. Equation 5.1 is valid if the template and profile are stable. For a profile to be stable its correlation coefficient with the template, $R = R(n)$, will improve according to $\langle 1 - R(n) \rangle \propto n^{-1}$ where n is the number of periods averaged over to make the template. The derivation of this result is given in Appendix E. In practise this is realised only after we have averaged some critical number of periods to make a profile. For smaller values of n, $\langle 1 - R \rangle$ will improve faster than n^{-1}. Breaks in $\langle 1 - R \rangle$ at certain values of n indicate periodic instabilities, e.g. drifting and nulling timescales, see Fig. 5.1. Beyond some value $n_{\rm crit}$, when $\langle 1 - R(n) \rangle \propto n^{-1}$ we say that $P(t)$ is stable. Figure 5.2 shows the behaviour of a stable pulsar. We note that it has, in the past, been mistakenly suggested that $\langle 1 - R(n) \rangle \propto n^{-0.5}$ signalled stability [10, 15, 22][1] but this is incorrect. For MSPs, this critical number of periods is $\lesssim 10^4$ and is always reached (as in Fig. 5.2) so that precision timing can be performed. In the case of slower pulsars the stability criterion is not reached [10, 22],

[1] Furthermore, in the past, arbitrary criteria for 'stability' have been set, e.g. Helfand et al. [10] defined stability as $R = 0.9995$.

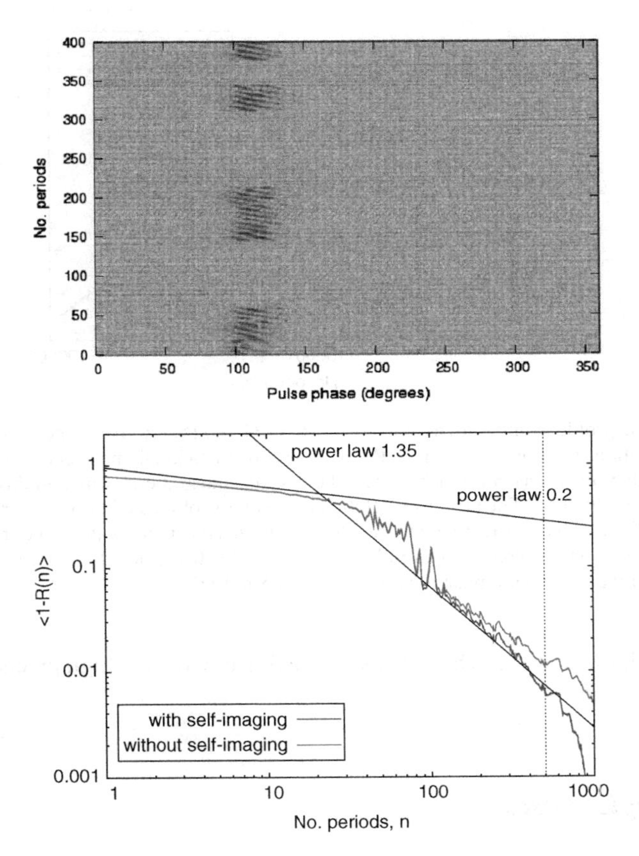

Fig. 5.1 The period-to-period variability of PSR B0031−07, a famous 'drifter' using data taken with the Westerbork Synthesis Radio Telescope (data kindly provided by B. W. Stappers and M. Serylak). (*Top*) A 'pulse stack' consisting of 400 pulse periods. The 'drift bands' can be clearly seen where the arrival time of the subpulses changes in phase periodically. Also evident are many nulls of differing lengths. (*Bottom*) A plot of $1 - R(n)$ where we can see two distinct power laws for different values of n. The *blue curve* suffers from 'self-imaging' as the template has been formed from the same dataset as have the sub-averages. The template used to produce the *pink curve* is a so-called 'noise-free' template and we can see that the self-imaging effect is removed. Helfand et al. [10] did not account for self-imaging and we can see their derived power law of 1.35 agrees with our self-imaged value but is steeper than the true value. It diverges even more beyond the 'half-template' mark (*the vertical dotted line*) which they did not consider. The break in the curve at ~20 periods is due to drifting and nulling timescales and the peaks seen in the range ~40−100 periods indicate nulling lengths. Clearly the n^{-1} regime has not been reached so much longer observations would be needed to perform precision pulsar timing of this source

nor is the precision as high given that the TOA error $\sigma_{\mathrm{TOA}} \propto W^{3/2} P^{-1/2}$ is larger for slow pulsars than for MSPs, where W and P are the pulse width and period, respectively. Furthermore the slower pulsars are observed to exhibit more glitches and more

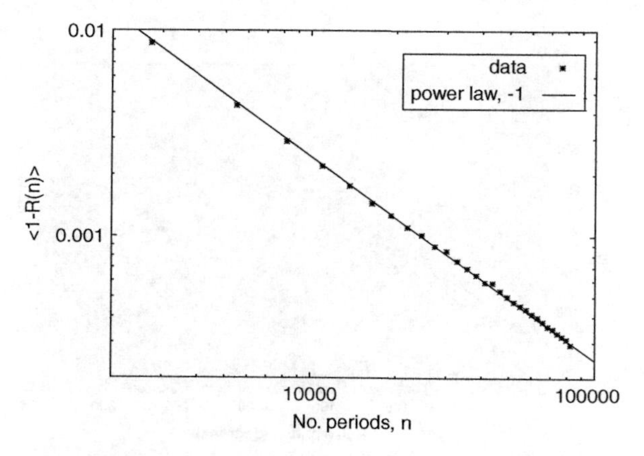

Fig. 5.2 Pulse profile stability assessment of PSR J0437−4715, the 'best timer' currently known, using observations taken with the Parkes Radio Telescope (data kindly provided by K. Liu). We can see that the stability criterion is reached from the first sub-integrations, which in this case are 2740 periods, 16 s, in duration. The power law plotted behaviour is obvious, implying our assumption of Eq. 5.1 is valid, i.e. that our template is excellent. This analysis ignores jitter and in fact it is jitter which is the dominant source of error in PSR J0437−4715. It is studied intensely for this reason, as its proximity and strength make it a prototype 'SKA pulsar'

so-called 'timing noise'.[2] Thus MSPs can be timed with very high precision whereas slow pulsars cannot.

5.1.2 Single Pulses

RRATs are generally detected via their sporadic single pulses as (by definition) they are only, or more easily, detectable in this way as opposed to frequency domain techniques. Their pulses are not detectable every rotation period with typical observed pulse-to-pulse separations ranging from $\sim$10 to $\sim$1000 periods so, unlike typical pulsars, we do not see strong pulse profiles when folding. This means we lose the two advantages of phase folding—stable profiles and increased signal-to-noise ratio. However the single pulses themselves are quite strong with typical peak flux densities of $\sim 10^2$–10^3 mJy (see Table 4.4). For the observations reported here the typical signal-to-noise ratios this corresponds to range from as low as 6 to as high as 60 so that, from a signal intensity point of view, timing RRATs from their single pulses is possible. However, the single pulse profiles are far from stable in phase. Phase stability is usually implicitly assumed (in timing analysis software) when using high S/N profiles and templates. This assumption is inappropriate for single-pulse timing and will result in extra scatter in our timing residuals with a magnitude given by the

---

[2] Timing noise is a red noise feature seen in pulsar timing residuals which may be related to pulsars switching between two spin-down rates [16].

size of the phase window wherein we see single pulses. As we will show this effect is clear in our data.

5.1.3 From Bits to BATs

01100010011010010111010001110011 →

Here we outline the steps involved in progressing from a telescope signal to barycentred pulse arrival times and a coherent timing solution.

(i) *Observe sources in 'search mode'.* As described in Sect. 4.1, filterbank data are taken utilising a bandwidth of ~250–300 MHz divided into ~500 channels, with 0.1–1.0 ms time sampling. The telescope receives dual-polarisations (linear at Parkes, circular at Jodrell Bank) but these are summed to produce total intensity, i.e. Stokes I. The data are 1-bit digitised before being written to tape. This time-frequency-amplitude data cube of 1 and 0 s is our raw data file. The exact observing specifications vary with telescope (Lovell or Parkes), the particular source of interest and other considerations such as the local RFI environment. The beginning of the observation is time-stamped according to the observatory clock. The time stamp, T_0 is known to 12 decimal places[3] in MJD (~80 ns). The time of the nth time sample is then simply $T_0 + (n \times t_{\mathrm{samp}})$.

(ii) *Search the data for single pulses.* The data is 'searched' for single pulses over a large DM range. This is to discriminate real pulses, which have a well-known shape as a function of DM [6], from terrestrial RFI signals. Such RFI 'pulses' peak at a DM of zero but can be strong enough to be detectable at very high DM values. The results of the single pulse search are inspected by eye and RRAT pulses are identified above a specified signal-to-noise ratio cut-off (which is typically at 5–6 σ but varies depending on the length of the observation, the presence or absence of RFI etc.). If several pulses are detected in an observation an additional check can be made to check whether pulses are in phase with each other, given the RRAT period (if known).

(iii) *Extract single pulse profiles.* Pulse profiles, centred on the pulse, and one period in length, are extracted at the known DM of the RRAT. This is done using a routine named rrat_prof, which writes ASCII profiles[4] known as 'mel files' which can be viewed and operated on within PSRPROF.[5]

(iv) *Obtain TOAs.* The profiles are cross-correlated with the template and the peak of the cross-correlation curve is deemed to be the TOA at the telescope, i.e. the

---

[3] This is the case for the hardware setup used in the observations described here, not a general rule for pulsar observations.

[4] These ASCII format mel files can be easily converted to other commonly used formats such as SIGPROC and EPN with existing routines within PSRPROF.

[5] http://www.jb.man.ac.uk/~pulsar/observing/progs/psrprof.html

site arrival time (SAT), which is referenced to the time stamp T_0. The templates used here are empirical and derived from smoothing each source's strongest observed pulse which results in simple one component templates. Averaging all of the (detected) individual pulses gives a wider pulse profile unsuitable for cross-correlating with individual pulses.

(v) *Convert SATs to BATs*. SATs, determined as above, are measured in Coordinated Universal Time (UTC). These are converted to barycentric arrival times (BATs), i.e. arrival times at the solar system barycentre at infinite frequency (with dispersion removed) in either Barycentric Dynamical Time (TDB) or Barycentric Coordinate Time (TCB). The definitions of and relationships between these time standards as well as all of the conversion steps are explained in detail in Appendix E.

Once we have obtained BATs we can model the timing parameters of the source. This is done using PSRTIME.[6] Expressing the rotational frequency of the pulsar as a Taylor expansion

$$\nu(t) = \nu_0 + \dot{\nu}_0(t - t_0) + \frac{1}{2}\ddot{\nu}_0(t - t_0)^2 + \cdots \tag{5.2}$$

the rotational phase (simply the integral of frequency with respect to time, modulo 2π) is given by

$$\phi(t) = \left[\phi_0 + \nu_0(t - t_0) + \frac{1}{2}\dot{\nu}_0(t - t_0)^2 + \frac{1}{6}\ddot{\nu}_0(t - t_0)^3 + \cdots\right] \pmod{2\pi}. \tag{5.3}$$

In addition to these terms, binary effects should be added (however none of the sources discussed here have detected binary companions) and the *observed* phase will be different due to positional uncertainties. Initial discovery observations have poorly constrained source positions and these result in sinusoidal variations in the observed pulse phase. Timing consists of minimising the χ^2 of the residuals of our timing model, i.e. the difference between our model for when pulses arrive and when they actually arrive.

Immediately after discovering and confirming a new source we know very little about it. If the rate of pulses is too low then we will not be able to determine an estimate of the period using period differencing (as in Sect. 4.4). If the rate is this low there is no way to proceed with timing the source. Assuming the rate is sufficient then we have an initial guess for the period and a knowledge of the sky position (uncertain to ~7 arcmin in both right ascension and declination, a PMPS beamwidth) which serves as our initial guess of the timing ephemeris. We can see from Eq. 5.3 that different effects will become visible in our residuals over different timescales. On the shortest timescale all we need to worry about is the rotation period, $P = 1/\nu$. We need to kick-start our timing solution by obtaining several closely spaced 'timing points', (say) every 8 h over the space of a day or two. This is necessary to build a coherent

[6] http://www.jb.man.ac.uk/~pulsar/observing/progs/psrtime.html

solution on short timescales as our initial knowledge of the period is not sufficient to be able to combine in phase TOAs obtained a day apart. For instance a RRAT with a 5 s period and a period uncertainty of $\Delta P = 1$ ms will have a phase uncertainty of $\sqrt{(86400/5)} \times 1.0$ ms ~ 17 s after a day so that it will be impossible to coherently combine observations spaced in this way. To coherently combine two observations of a source we want $(\Delta P/P)T_{\text{gap}} \lesssim 0.5P$. For our example ($P = 5$ s, $\Delta P = 1$ ms) our initial observations should be spaced no more than 3 h apart to avoid 'losing a turn' of the star. Once this has been done the period will be known to sufficient accuracy that all our TOAs over the timescale of a few days will be in phase. If we monitor the source like this over several months we will notice a quadratic signature appear in our residuals. This is the effect of the frequency derivative $\dot{\nu}$ (which is initially set to zero). This $\dot{\nu}$ effect is seen over a timescale of weeks to months. Positional uncertainties result in sinusoids, with periods of 1 year, appearing in the residuals. It is impossible to disentangle the effects of spin-down rate and positional uncertainty until at least 6 months of monitoring has been made, and preferably at least one year (i.e. a quadratic is highly covariant with half a sine wave!). Below, and in Chap. 6, we report timing solutions consisting of ν, $\dot{\nu}$, right ascension and declination. Figure 5.3 shows the effects (on the timing residuals) of not fitting for any of these parameters.

5.2 Timing at Jodrell Bank

Timing observations of RRATs began in August 2008. All eight of the original eleven sources with sufficiently high declination[7] were observed with the 76-m Lovell Telescope. Of these sources only two were clearly detected—J1819−1458 and 1913+1333, the sources with the strongest reported peak flux densities [20]. A campaign of regular monitoring observations of these sources began from this point onwards. Timing with the Lovell Telescope has the advantage that it has a large amount of dedicated pulsar observing time so that it is possible to space observations as regularly as is needed to build up a timing solution. The is crucial but unfortunately not possible at Parkes.[8] The Jodrell observations of J1819−1458 showed many, strong pulses so that it was clear that timing it at Jodrell Bank would be effective. The observations even showed numerous instances of consecutive pulses from J1819−1458, which had not been reported previously. J1913+1333 was seen to be weaker, but nonetheless clearly detectable. Regular monitoring of this source was needed as it turned out that the originally published timing solution was slightly incorrect.

---

[7] The effective declination limit for Jodrell Bank is $-32°$ but to perform longer observations with low spillover even more northernly sources than this are needed.

[8] This is due to the much more restricted observing schedule at Parkes. It has a high over-subscription rate because of its ideal location (for Galactic observations) at latitude $-33°$. The time allocated to a given project at Parkes is highly competitive to obtain and inflexible to 'on-the-fly' changes.

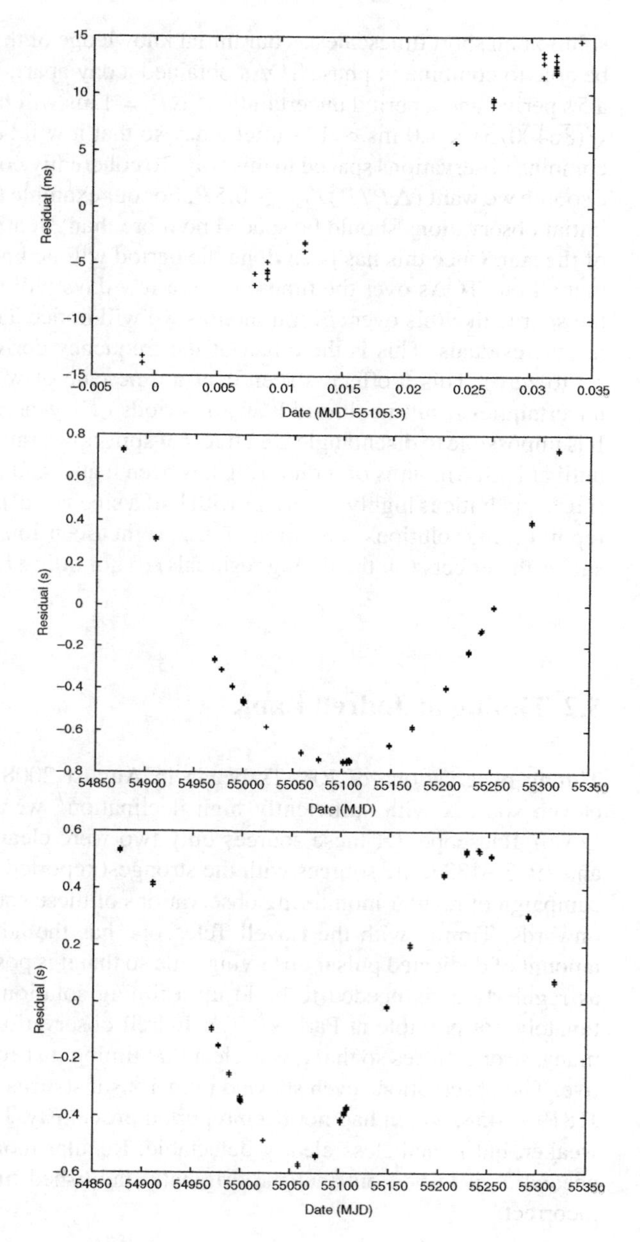

Fig. 5.3 Plotted are timing residuals for J1514−59 with the characteristic signatures of errors in frequency (*linear*), frequency derivative (*quadratic*) and position (*sinusoidal*) evident. *Note* the different timescales involved

Up to August 2009, observations at Jodrell were performed using an analogue filterbank (AFB) with 64 × 1-MHz channels centred at 1401 MHz and with a time sampling of 100 μs. Since August 2009 the newly acquired digital filterbank (DFB) has been used. This has increased the available bandwidth to 512 MHz, although in practise only about half of this is usable (see Sect. 5.2.4). The bandwidth is divided into 1024 $\times$ 0.5-MHz channels, reducing the effect of dispersion within channels. Initially time sampling of 250 μs was used but due to the increased data volume for DFB observations, this has been adjusted so that it is now more common to observe at a rate of 1 ms. The DFB allows the use of a large number of bits, however, due to data volumes and decreased effectiveness of RFI excision, 1-bit digitisation remains the norm.

5.2.1 Pulsar Glitches

Many pulsars have been observed to undergo glitches. As described in Sect. 2.2.2, these are step changes in spin frequency ν and its derivative $\dot{\nu}$, which are thought to be due to the transfer of angular momentum between the internal superfluid and the stellar crust. A glitch at $t = 0$ takes the form

$$\nu(t) \rightarrow \nu(t) + \Delta\nu_{\rm p} + \Delta\dot{\nu}_{\rm p}t + \Delta\nu_{\rm d}e^{-t/\tau_{\rm d}} \tag{5.4}$$

$$\dot{\nu}(t) \rightarrow \dot{\nu}(t) + \Delta\dot{\nu}_{\rm p} + \Delta\dot{\nu}_{\rm d}e^{-t/\tau_{\rm d}} \tag{5.5}$$

where the permanent steps are labelled with a 'p' and the steps labelled 'd' decay on a timescale of $\tau_{\rm d}$ [24]. Here $\nu(t)$ is as given in Eq. 5.2. Glitches sometimes dominate the long-term spin evolution of pulsars [18] and have been observed in numerous young pulsars and several magnetars, appearing to be a normal phenomenon among rotating neutron stars [9]. 'Glitch sizes', $\Delta\nu/\nu$, are commonly quoted and range from $\Delta\nu/\nu \sim 700 \times 10^{-6}$ seen in the Crab, down to the smallest glitches with $\Delta\nu/\nu \sim 10^{-11}$, which are very difficult to distinguish from the so-called 'timing noise' seen in the long-term timing residuals of many pulsars [11].

5.2.2 J1819−1458

RRAT J1819−1458 (1819 from herein) is easily detected with the Lovell Telescope at Jodrell Bank. In a one-hour observation it is common to detect 20 pulses with signal-to-noise ratios approaching 30. Since August 2008 regular timing observations, typically weekly, have been made at Jodrell Bank. Some observations of 1819 are still made at Parkes but these are at most monthly but typically less frequent. A number of discoveries have been made in our timing observations of 1819. These include: (1) It seems to have three (or possibly more) distinct sub-pulse 'windows' (ranges in pulse phase) wherein we see pulses, (2) It has undergone glitches with unusual properties not seen in any previously recorded glitches, and (3) There is an indication that it may show γ-ray emission in phase with its radio profile.

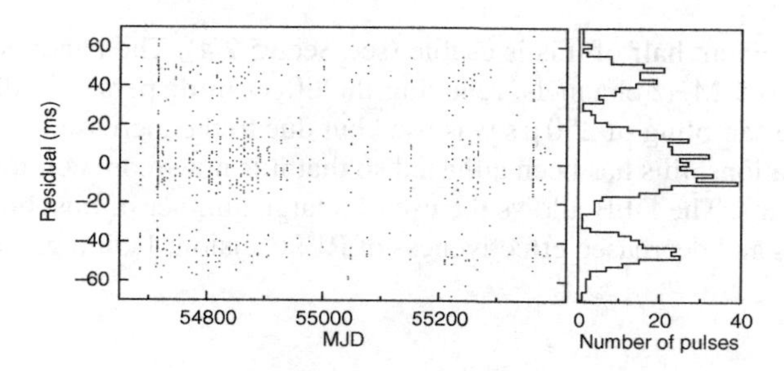

Fig. 5.4 (*Left*) Timing residuals for all pulses detected at Jodrell Bank between August 2008 and July 2010. The three bands are evident. (*Right*) A histogram of the residuals, which essentially gives a probability density distribution in pulse phase for 1819

Tri-Modal Residuals

Figure 5.4 shows the timing residuals for all TOAs recorded at Jodrell Bank for 1819 between August 2008 and July 2010. The three main sub-pulse regions are clearly visible. We can see that ~60% of pulses are in the central band. It is inappropriate to assume that the scatter in this plot is a good measure of our timing residuals, it is simply a consequence of timing sub-pulses with a probability distribution as given in Fig. 5.4 using a single-peaked template. The scatter within the bands is a better estimate of the true rms of the residuals so, as most TOAs are easily identified with one of these bands, we apply two 45-ms 'jumps' to our data in order to lower the top band and raise the bottom band. Aligning the bands like this decreases the rms of the residuals from 21.2 to 9.1 ms. The uncertainties in the fitted parameters are similarly reduced. This 'banding' is seen in both Lovell and Parkes TOAs [13] although observations at Urumqi show only two bands with very occasional pulses from the top band [8]. Our observations show that the pulses detected in the top and bottom bands have similar intensity distributions [17]. The explanation of this effect is unclear, but does not seem to be due to observing frequency, e.g. the bands having different spectral indices, as the bandwidths covered are essentially identical. However, the Urumqi observations were only sensitive to bursts with flux densities above 3.4 Jy (yielding 162 pulses in 94 h of observation) and most of the pulses detected at Jodrell Bank were weaker than this, so that given the unknown amplitude distribution at high flux density, the two sets of results appear consistent.

Anomalous Glitches

Timing 1819 in this way we have been able to coherently connect the Jodrell TOAs to the earlier Parkes TOAs. The third, newly identified (see Chap. 4) detection of 1819 in the PMPS, has helped us to be able to coherently connect back as far as the original survey pointings. These have enabled us to identify two glitches which are shown in Fig. 5.5. The first glitch is the largest with a fractional size $\Delta\nu/\nu \approx 700 \times 10^{-9}$

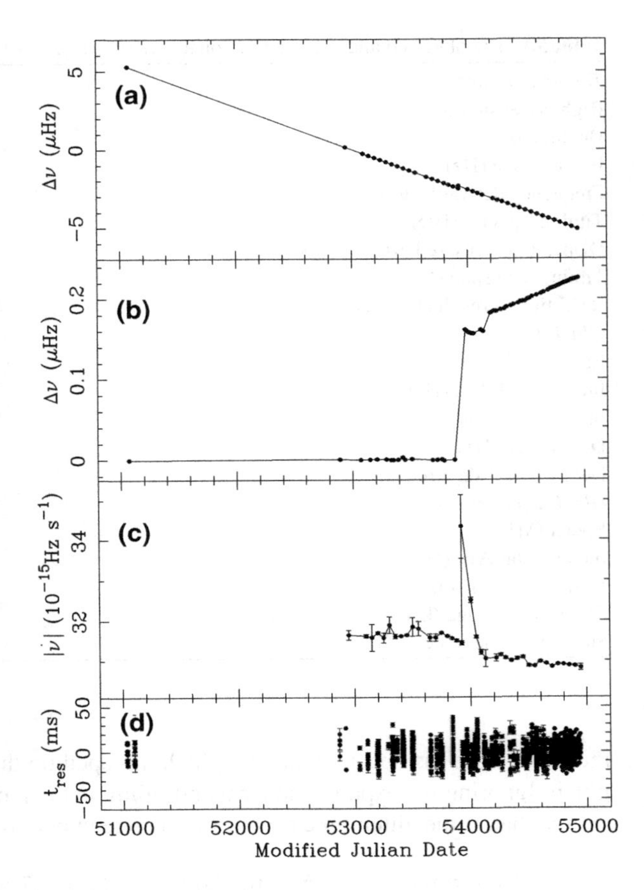

Fig. 5.5 The evolution of rotation frequency of 1819 over the last decade. The data points just after MJD 51000 are the three PMPS detections. **a** Shows the secular slowdown in ν. A discontinuity, the signature of a glitch, can be seen at MJD 53900. **b** With the average slope subtracted, and with the data fitted between MJD 5100 and 53900 with a simple slow-down model, we can clearly see the first glitch as well as a second smaller glitch, about 200 days later. **c** Presents the variation in the magnitude of $|\dot{\nu}|$, showing a significant decrease in the rate of slowdown following the glitches. **d** shows the timing residuals $t_{\rm res}$ relative to the rotational model given in Table 5.1

whereas the second glitch is smaller with $\Delta\nu/\nu \approx 70 \times 10^{-9}$. These are similar in size to the glitches seen in young pulsars [28] and in magnetars [21]. The fact that we can connect back to the PMPS pointings means that 1819 did not suffer any large glitches in the gap between the survey and first follow up observations, although small glitches cannot be ruled out. The fitted glitch parameters are shown in Table 5.1.

The most noteworthy point is that the net effect of the glitches is so as to decrease the magnitude of the slow-down rate ($|\dot{\nu}|$ and $\dot{P}$ lower, $\dot{\nu}$ higher) of the star's rotation. This is completely anomalous and unlike all radio pulsar glitches ever detected. Figure 5.6 illustrates this net effect on $\dot{\nu}$ for 1819 as well as for a sample of glitches in other pulsars. What does this mean for the spin evolution of 1819? As the star spins it slows down according to a spin-down law. Let us assume that this is of the form $\dot{P} = KP^{2-n}$, Eq. 2.22 in terms of spin periods. The glitches represent instantaneous jumps added to this evolution in $P - \dot{P}$ space. The change in period is $\Delta P = -\Delta\nu P^2$. From Eq. 5.4 we can see that $\Delta\nu = \Delta\nu_{\rm p} + \Delta\dot{\nu}_{\rm p}t + \Delta\nu_{\rm d}e^{-t/\tau_{\rm d}}$. From Table 5.1 we can see that all the terms in this expression are positive so that

Table 5.1 The observed and derived rotational parameters of 1819

Timing parameters	
Right ascension α	$18^{h}19^{m}34^{s}.173$
Declination δ	$-14°58'03''.57$
Frequency ν (Hz)	0.23456756350(2)
Frequency derivative $\dot{\nu}$ (s^{-2})	$-31.647(1) \times 10^{-15}$
Timing Epoch (MJD)	53351.0
Dispersion measure DM (cm^{-3} pc)	196
Timing data span (MJD)	51031–54938
RMS timing residual σ (ms)	10.2
Glitch 1 parameters	
Epoch (MJD)	53924.79(15)
Incremental $\Delta\nu_p$ (Hz)	$0.1380(6) \times 10^{-6}$
Incremental $\Delta\dot{\nu}_p$ (s^{-2})	$0.789(6) \times 10^{-15}$
Decay $\Delta\nu_d$ (Hz)	$0.0260(8) \times 10^{-6}$
Decay timescale τ (days)	167(6)
Glitch 2 parameters	
Epoch (MJD)	54168.6(8)
Incremental $\Delta\nu_p$ (Hz)	$0.0226(3) \times 10^{-6}$
Derived parameters	
Characteristic age (kyr)	120
Surface magnetic field (G)	50×10^{12}

the change in P is negative, i.e. the glitch has sped up the star, as expected from a glitch. Ignoring the exponential term and filling in the numbers for the large glitch, we see that, if the glitch time $t = t_0$, the change in period is:

$$\Delta P(t - t_0)_{1819} = -2.5 \times 10^{-6} H(t - t_0)[1 + 5.7 \times 10^{-9}(t - t_0)], \tag{5.6}$$

where $H(t - t_0)$ is the Heaviside function. The first term represents the instantaneous change. The second term is due to the change in rotation rate, it is zero at $t = t_0$, and is only comparable to the first term after ~2000 days. The fractional change in period is small with $\Delta P/P \approx 5 \times 10^{-7}$. The change in period derivative is $\Delta\dot{P} = -\Delta\dot{\nu}P^2 - 2P\Delta P\dot{\nu}$. Substituting $\Delta\dot{\nu}$ from Eq. 5.5 and inserting the numbers we get

$$\Delta\dot{P}(t - t_0)_{1819} = -1.43 \times 10^{-14} H(t - t_0)[1 + 19(-\Delta P)]. \tag{5.7}$$

The first term dominates here and we can see that the fractional jump in period derivative is appreciable with $\Delta\dot{P}/\dot{P} \approx -0.02$. From Eq. 2.21 we know that $\Delta B/B = 1/2(\Delta P/P + \Delta\dot{P}/\dot{P})$ so that the glitches have resulted in a 1% step decrease in surface magnetic field strength. This is in stark contrast to all other recorded glitches which result in an increase in B.

But how are we to interpret these changes? When contemplating the significance of this effect we might consider that the effect of the glitches in $P - \dot{P}$ space is to move

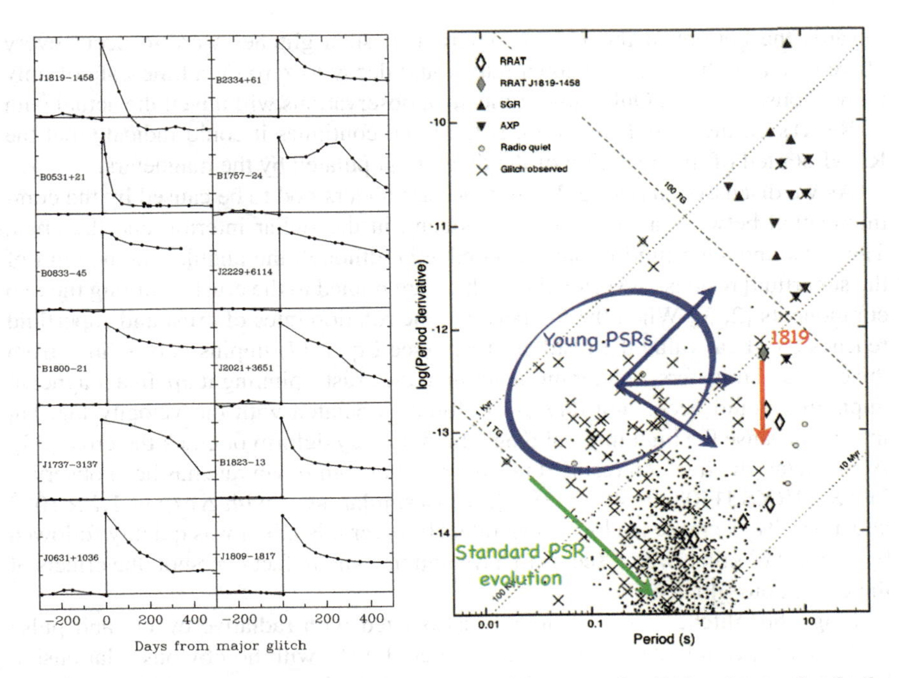

Fig. 5.6 (*Left*) The relative change in the magnitude of $\dot{\nu}$ is shown for 1819 and a sample of pulsar glitches, for a few hundred days around typical glitches, from the Jodrell Bank Glitch Database [9]. The behaviour of the Crab pulsar (B0531+21) shows the greatest similarity to that of 1819, each displaying a relatively short-term transient before reaching a new asymptotic slow-down rate, although the long-term net increments in slow-down rate clearly have different signs. (*Right*) The top half of the $P - \dot{P}$ diagram indicating standard dipole pulsar evolution (*green arrow*) according to Eq. 2.22, the direction of observed evolutionary paths (*blue arrows*) in young pulsars, and the direction of movement of 1819 (*red arrow*) due to the anomalous glitches. The arrows shown are indicative only, their lengths are arbitrary

J1819−1458 downwards (see Fig. 5.6). If we were to propose that such glitches were typical in this source then it would suggest that J1819−1458 previously occupied the region of $P - \dot{P}$ space where the magnetars are. The importance of such effects must be considered when we consider pulsar (and magnetar) spin evolution in the $P - \dot{P}$ diagram. Objects with constant dipolar magnetic field move towards the lower right-hand corner of the diagram with a slope of −1. To measure such motion we need to measure $\ddot{P}$, something which is only possible if $\dot{P}$ is high, i.e. in the 'young' pulsars. These are seen to move with a slope of between −1.0 and +0.5. Note that the slope is related to the braking index n by slope $= 2 - n$ so that these slopes correspond to braking indices from 3 to 1. Glitches observed in these normal pulsars, like those shown in Fig. 5.6, result in a net increase in slow-down rate and an upwards step in the $P - \dot{P}$ diagram. On the other hand, RRAT J1819−1458 has stepped vertically downwards, towards smaller values of $\dot{P}$. If this particular post-glitch behaviour is typical, then the long-term effect of any glitches would be a secular movement towards the bottom of the $P - \dot{P}$ diagram. If such glitches were to occur every 30 years (say), then the slowdown rate would decay to zero on a timescale of only a few thousand years. Only larger time-span observations will unveil the actual path of RRATs on the $P - \dot{P}$ diagram. If the trend continues it could indicate that the RRAT started off in the region of the diagram populated by the magnetars.

As we discussed in Sect. 2.2.2, glitches are understood to be caused by the communication between a superfluid component, in the stellar interior, and the crust. The crust and superfluid rotate independently although the angular momentum of the superfluid resides on vortex lines which are pinned to the crust, coupling the two components [2, 3]. When the lag between the rotation rates of crust and superfluid reaches a critical value the resultant force (see Eq. 2.17) unpins vortex lines from the crust and transfers angular momentum to the crust, spinning it up. In a magnetar, unpinning could occur, not due to the force associated with the velocity lag, but instead because the high internal magnetic field may deform or crack the crust [25]. A magnetar glitch which resulted in a decrease in spin-down rate has been observed in AXP 1RXS J170849−400910 [12], with a similar 'size' with $\Delta\nu/\nu = 1.2 \times 10^{-6}$ and a $\sim$0.7% decrease in slow-down rate. However this glitch was quickly followed by a 'normal' glitch which seems to have undone, and in fact overshot, the effects of the anomalous glitch on $\dot{\nu}$.

Magnetar glitches are occasionally associated with radiative events and pulse profile and spectral changes (see, e.g. Dib et al. [7]), with no obvious relationship between the size of the glitch and the extent of these changes. In 1819 we see an increase in both burst rate and peak pulse energy immediately following the first glitch (see Fig. 5.7), which is suggestive of an association, although this may be a statistical fluke. More large glitches must be detected for such correlations to be tested robustly. In addition, X-ray observations immediately after the next large glitch would be useful to search for magnetar-like bursts or correlated pulse profile or spectral changes.

Fermi Detection?

The Fermi γ-ray satellite was launched on the 11th of June 2008 and since then has collected the highest quality γ-ray data ever taken using the onboard Large Area Telescope [4]. Fermi can perform blind searches for pulsars (in a very similar manner to the pulse differencing technique used for PMSingle source, as described in Sect. 4.4), but typically folds on known radio ephemerides, which are maintained by a network of collaborating radio pulsar observers, which include researchers at Jodrell Bank. Fermi looks for γ-ray emission from all pulsars with $\dot{E} > 10^{33}$ ergs s^{-1} and a random selection of other sources across the $P - \dot{P}$ plane [1]. J1819−1458, which has $\dot{E} = 3 \times 10^{32}$ ergs s^{-1}, is not part of the Fermi timing list, although, our timing observations of 1819 at Jodrell Bank, conveniently, cover the same period as the operation of Fermi. Using our best radio timing model over this time we have folded the γ-ray photons received by Fermi according to the ephemeris of 1819. We have taken an aperture on the sky of 1° and taken into account the energy dependent

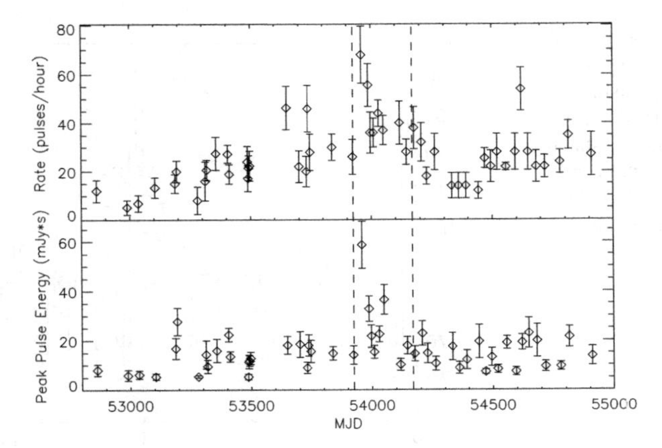

Fig. 5.7 The burst rate (*top*) and average pulse energy (*bottom*) for 1819, as a function of time. The times of the two glitches are marked with vertical dashed lines. For both quantities, there is a large variation, with maximum values in the observation (on MJD 53960) which immediately follows the first, and largest, glitch. On this date, the mean pulse detection rate is 68 ± 12 pulses/hour, 2.8 times the mean and significant at the 3.5σ level. The corresponding value of the peak pulse energy is 58.7 mJy s, which is 3.7 times the mean and significant at the 4.7σ level. We note, however, that the high burst rate on MJD 54619 is not associated with any obvious timing abnormality

point-spread function of Fermi as well as γ-rays from the Earth. The resultant γ-ray profile is shown in Fig. 5.8. The profile seems to show a peak which is in phase with the radio peak (defined as a phase of 1), which differs from what is seen for the majority of pulsars with Fermi, where two γ-ray peaks, out of phase with the radio are typically seen [1]. However, the significance of the peak is low ($\lesssim 4\sigma$) and it seems only to be present in the second half of the Fermi dataset. As Fermi releases new data we can test this significance. The radio TOAs have been cross-correlated with γ-ray photon arrival times and no significant signal is seen. We note that the Fermi data has similarly been folded on the ephemerides of several other RRATs with timing solutions (those presented in Chap. 6) with no significant detections.

5.2.3 J1913+1333

Although J1913+1333 (1913 from herein) had a coherent timing solution published in the original discovery paper [20], after 18 months of monitoring the TOAs began to deviate from it, i.e. the initial solution was slightly in error. We began timing 1913 at Jodrell Bank in August 2008 and have converged on a stable solution. The timing residuals over this time are shown in Fig. 5.9. 1913 is not as easily detected with the Lovell Telescope as 1819. In a two-hour observation, with the AFB it was typical to detect anything from 0 to 10 pulses with signal-to-noise ratios usually less than 15. With the DFB, there has been one half-hour observation which has

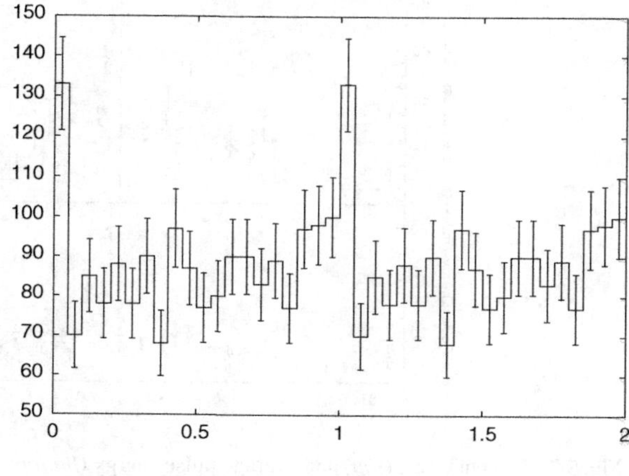

Fig. 5.8 The γ-ray profile of 1819 folding data from the Fermi telescope using the radio ephemeris determined at Jodrell Bank. As stated in the text, this signal is only evident in half of the Fermi dataset. (Image credit: P. Weltevrede)

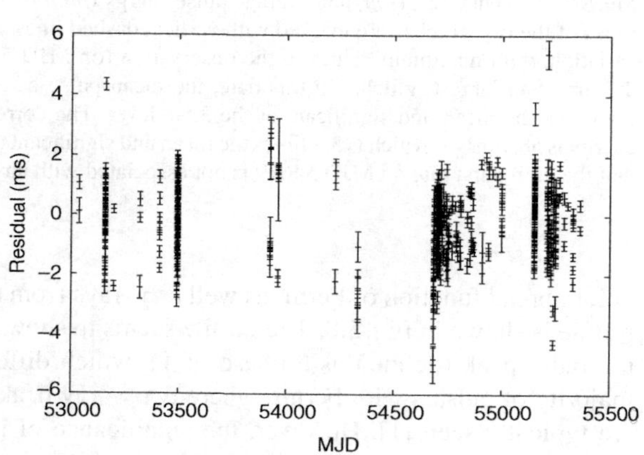

Fig. 5.9 The timing residuals for 1913 for the last ~2500 days. The Jodrell Bank TOAs are those more densely spaced from ~54700 onwards. The initial timing solution was determined from the more sparsely spaced Parkes TOAs before this point

detected 36 pulses as well as several non-detections. Clearly the source is highly variable. Pulses, when they are detected, are usually seen to come in clumps, i.e. the likelihood of detecting a pulse is higher immediately after detecting a pulse. The new timing solution changes $\dot{P}$ and the position, which can be somewhat covariant, as we can see in Fig. 5.3. The $\dot{P}$ increases by 10% from 7.9 to 8.7×10^{15}. The change in right ascension is small at ~4 arcsec but the change in declination is considerable at 2.8 arcmin. Although this may not seem like much of a change in comparison to the resolution of single-dish radio telescopes, it is significant compared to the positional accuracies required for high-energy observations. This has led to X-ray and infrared observations of the wrong position: 9.3 ks (2.5 h) with *Swift*-XRT [19] and ~4 h with the VLT, respectively [23].

Table 5.2 The specifications of the Analogue filterbank (AFB) and Digital filterbank (DFB) observations of RRATs at Jodrell Bank

	Analogue filterbank	Digital filterbank
Centre frequency (MHz)	1402.5	1524.75
Bandwidth (MHz)	64	512
Usable bandwidth (MHz)	64	$\lesssim$250
Frequency channels	64	1024
Time sampling	100 μs	1 ms
Number of bits	1	2
File size	0.14 GB/30 min	0.44 GB/30 min

5.2.4 The JBO DFB

In January 2009, a new digital filterbank (DFB) was acquired for the Lovell Telescope. After a few months of technical work it was operational and standard (i.e. folded) pulsar observations were made using the new system. Due to a large number of teething problems, it was not until August 2009 when the first stable 'search mode' data was recorded using the DFB. From then onwards, RRAT observations were carried out using the DFB in this mode. Table 5.2 gives the specifications of the DFB system as contrasted to the previously used AFB system. We can see that the transition constitutes a huge improvement in observing capabilities, but, with great power comes great responsibility, and many new problems, bugs and 'undocumented features' have made observing with the DFB a challenge. Here we describe the DFB observations of RRATs and highlight some of the main associated difficulties.

Immediately we can see that the DFB affords us much more bandwidth and frequency resolution. We can also avail of a high dynamic range and digitise our data with a large dynamic range, if desired. It was decided to use many narrow frequency channels to reduce dispersive smearing within channels and to allow removal of narrowband interference without sacrificing any clean parts of the band. In switching from AFB to DFB observations we can see that our data files (for the same observing time) will increase in size by $(1024/64)n_{\rm bits}$ for $n_{\rm bits}$ sampling and the same time resolution. For fast sampling with large dynamic range this can make data files impractically large for processing. After some experimentation it was decided to use 2-bit sampling and a slower time sampling of 1 ms while keeping the full frequency resolution. Such time sampling is ample for observations of RRATs whose pulses are on the order of a few milliseconds. These specifications give a data rate of $\sim$1 GB per hour. The increased bandwidth should increase sensitivity, which, in theory, may enable shorter observations than with the AFB. However the large bandwidth adds many problems. Firstly, only $\sim$50% of the band is usable due to strong terrestrial signals (mobile phone signals, TV, radar, GPS, etc.) within the band. Figure 5.10 shows a bandpass for a typical DFB observation. The two 'humps' indicate frequencies which are essentially always unusable and are masked/removed for post-processing.

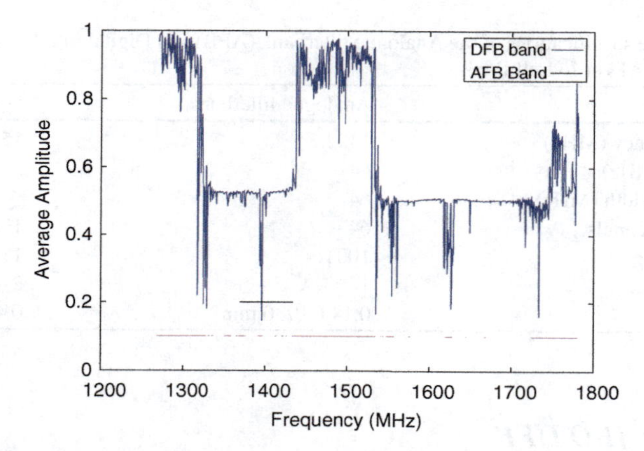

Fig. 5.10 The bandpass for the first DFB 'search mode' observation of a RRAT. The two 'humps' are unusable frequencies and must be masked/removed when processing the data, i.e. anything involving 'frequency-scrunching', i.e. adding frequency channels with ($DM > 0$) or without ($DM = 0$) appropriate delays. As we can see this observation was 1-bit digitised: an ideal bandpass would have all channels with average value of 0.5, i.e. half 1s, half 0 s

The extended bandwidth has also resulted in radar signals being detected. These originate from Clee Hill Radar Station which is used as part of National Air Traffic Services and monitors air traffic within a 160 km (100 mile) radius. Unfortunately, Clee Hill is a mere 93 km (58 miles) from Jodrell Bank so that the radar signal is very strong, and, in many senses, it is very well-designed RFI, which is difficult to excise. The radar signal is pulsed, with $\sim$100 μs pulses emitted every $\sim$3.4 ms. The time between pulses is varied so that the reflected signal can be used to infer unambiguous distances. The radar transmitter rotates at a rate of approximately once every 8 s. This rotation rate varies however, at the level of several milliseconds, most likely due to the accuracy limits of the mechanical structure of the transmitter. As the transmitter rotates, it cuts our line of sight for $\sim$50 ms. These pulses impart much structure in the fluctuation spectrum of our signal. As the pulses are narrow we get many harmonics.[9] It is not uncommon for several hundred harmonics to be observed at $n \times 0.125$ Hz, denoting the rotation rate of the transmitter. More strong Fourier domain spikes are seen at a frequency of $\sim$290 Hz, and harmonics of this. This is the rate at which pulses are emitted. They are broadened in frequency due to the changing time between pulse emission. Also, around this frequency, in an envelope of $\sim$20 Hz, the inverse of the time the radar is in our line of sight, we see more harmonics of the 0.125 Hz signal. Considering that the 8 s rate varies, it can only be determined to at best $\sim$0.1 ms for a typical observation and it has hundreds of harmonics, we can see that it is very difficult to remove all of the radar induced structure in the frequency domain. Figure 5.11 shows a typical radar detection.

[9] A good rule of thumb is that the number of harmonics you expect in the frequency spectrum is roughly the period-to-width ratio of the pulses [15].

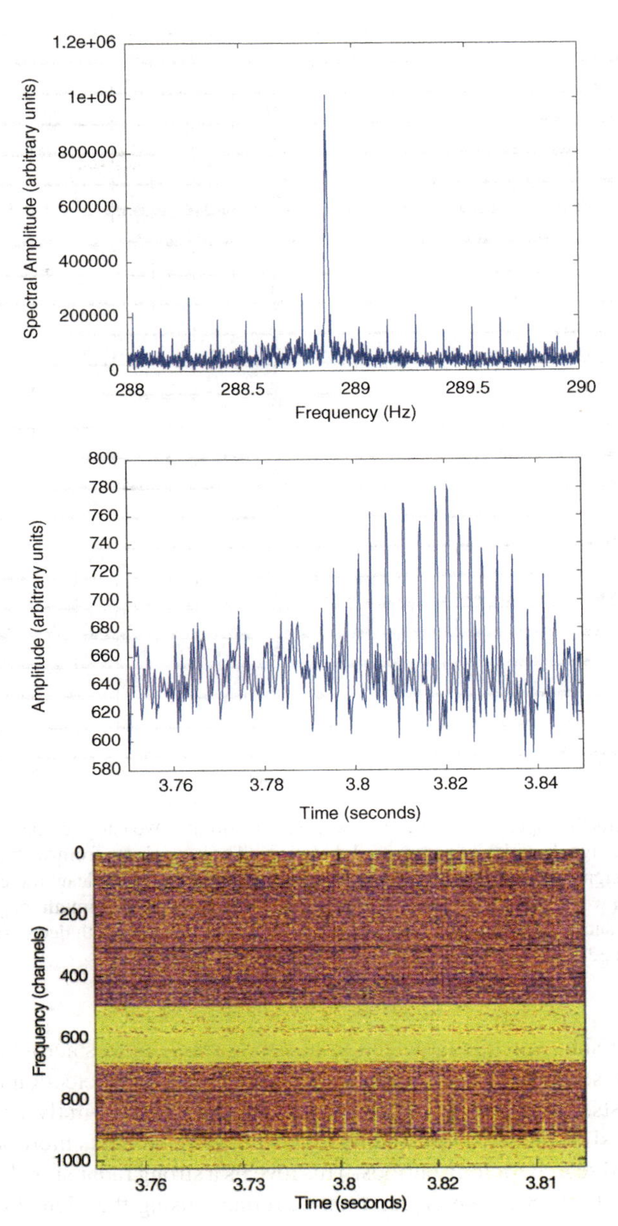

Fig. 5.11 (*Top*) The 3.4 ms feature in the frequency domain. Within an envelope of $\pm\sim 10$ Hz of this feature we see more harmonics (the ∼2300th harmonic!) of the 8 s rotation rate. (*Middle*) A series of detected radar pulses in the time domain at $DM = 0$. (*Bottom*) The same radar pulses, shown as a function of observing frequency. The horizontal bands correspond to the bad 'humps' in Fig. 5.10. This plot shows raw filterbank 1-bit data, yellow denotes 1 and purple denotes 0

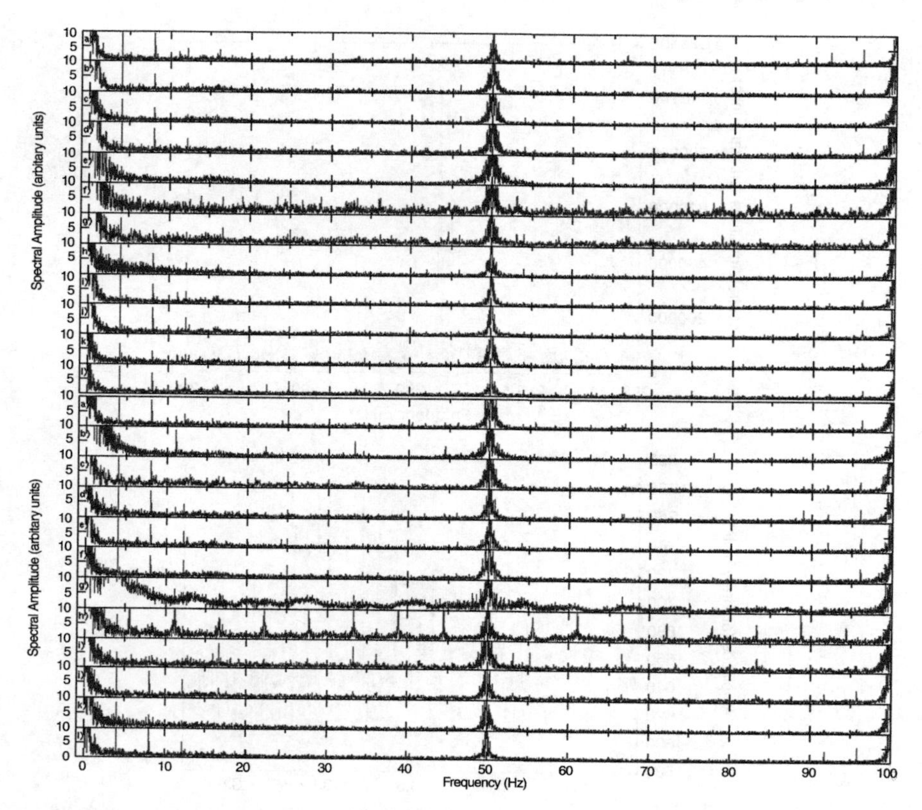

Fig. 5.12 Fluctuation spectra for 24 × 15° sections of azimuth. We can see that there are several strong periodic signals which are seen in all directions. These originate from within the observing system. Other signals are clearly directional. Unfortunately there are no 'clean' directions, although some are much worse than others. The top spectrum corresponds to an azimuth range from 320° to 305° and each successive spectrum corresponds to a 15° wedge in azimuth decreasing in azimuth, i.e. going from 320° → W → S → E → N → 320°

This radar signal results in detection of strong single pulses across a wide range of DM (up to several 100 cm^{-3} pc is typical) thus hindering detection of true astrophysical bursts. We endeavour to remove it in several ways. Firstly we identify the period of the signal using a fast folding algorithm [14], which is more sensitive than an FFT period search for long periods. This reveals a strong radar signal with a period in the range of 7.9–8.1 s which can then be refined, using the PDMP tool from the PSRCHIVE suite of packages,to give a period accurate to ∼0.1 ms. It is difficult to accurately replace the samples, wherein we see radar, with noise, and attempting to do this usually produces 'edges' which are detected as pulses.[10] Instead we can just

[10] We note that this would not be a problem for periodicity searches, and such procedure is being used to great effect in the analysis of the High Time Resolution Universe Survey at present.

mask the times of the radar in our single pulse search. An option was created for the search tool destroy to take an offset, period and window size for masking, and these sections of the time series are ignored in calculating the time series mean and rms and in threshold searching. Another method for removing periodic signals is to zap spikes in the fluctuation spectrum and this is also employed when appropriate.

Unfortunately, this known radar signal is not all that is to be dealt with in terms of RFI sources. A test observation was performed using the Lovell Telescope at Jodrell Bank in April 2010 (after the Galaxy had set) to determine the directionality of RFI signals. The observation was made while the telescope was slewing through a full 360° in azimuth at a constant elevation of 5°. A full rotation takes 24 min and the data were recorded using a time sampling of 200 μs, a bandwidth of 512 $\times$ 1 MHz channels, 4-bit digitisation and with one polarisation product (i.e. total intensity, Stokes I). To examine the directionality, the data were divided into 1 min segments, each representing 15° in azimuth, and fluctuation spectra were determined. These are shown in Fig. 5.12. We can see that all directions contain strong periodic signals—both non-directional (e.g. the 50 Hz mains frequency and an, as yet, unidentified 4 Hz source) and directional. In the azimuth range of 245°–215° we see many RFI features. Clee Hill is at a heading of 192° although at the resolution displayed here the 8 s harmonics are not visible. The worse spectra are seen in the range 5°–50°, but several directions show a strong RFI, in particular a strong 11 Hz signal with many harmonics is evident in the range 20°–35°. Jodrell Bank is a mere 16 km (10 miles) south of Manchester airport which is within this azimuth range, as is the city of Manchester. While in theory we might restrict the azimuth ranges wherein we observe, for a number of reasons this is not practical—it is difficult to schedule, the Galactic plane is in the south (!) and RFI signals can enter our receiver regardless of whether we are pointed at them or not, i.e. we will always be stuck with some amount of RFI and it is our job to limit this amount by whatever practical means we can think of.

References

1. A.A. Abdo et al., ApJS **187**, 460 (2010)
2. M.A. Alpar, D. Pines, in *Isolated Pulsars*, vol. 1, ed. by K.A. van Riper, R. Epstein, C. Ho, Isolated Pulsars (Cambridge University Press, 1993), p. 17
3. P.W. Anderson, N. Itoh, Nature **256**, 25 (1975)
4. W.B. Atwood et al., ApJ **697**, 1071 (2009)
5. N. Bartel, D. Morris, W. Sieber, T.H. Hankins, ApJ **258**, 776 (1982)
6. J.M. Cordes, M.A. McLaughlin, ApJ **596**, 1142 (2003)
7. R. Dib, V.M. Kaspi, F.P. Gavriil, ApJ **673**, 1044 (2008)
8. A. Esamdin, C.S. Zhao, Y. Yan, N. Wang, H. Nizamidin, Z.Y. Liu, MNRAS **389**, 1399 (2008)
9. C.M. Espinoza, Ph.D. thesis, University of Manchester, 2009
10. D.J. Helfand, R.N. Manchester, J.H. Taylor, ApJ **198**, 661 (1975)
11. G. Hobbs, A.G. Lyne, M. Kramer, C.E. Martin, C. Jordan, MNRAS **353**, 1311 (2004)
12. G.L. Israel, D. Götz, S. Zane, S. Dall'Osso, N. Rea, L. Stella, A&A **476**, L9 (2007)

13. A. Karastergiou, A.W. Hotan, W. van Straten, M.A. McLaughlin, S.M. Ord, MNRAS **396**, 95 (2009). astro-ph/0905.1250
14. V.I. Kondratiev, M.A. McLaughlin, D.R. Lorimer, M. Burgay, A. Possenti, R. Turolla, S.B. Popov, S. Zane, ApJ **702**, 692 (2009)
15. D.R. Lorimer, M. Kramer, *Handbook of Pulsar Astronomy* (Cambridge University Press, 2005)
16. A. Lyne, G. Hobbs, M. Kramer, I. Stairs, B. Stappers, Science **329**, 408 (2010)
17. A.G. Lyne, M.A. McLaughlin, E.F. Keane, M. Kramer, C.M. Espinoza, B.W. Stappers, N.T. Palliyaguru, J. Miller, MNRAS **400**, 1439 (2009)
18. A.G. Lyne, R.S. Pritchard, F. Graham-Smith, F. Camilo, Nature **381**, 497 (1996)
19. M. McLaughlin, in *Astrophysics and Space Science Library*, Vol. 357, ed. by W. Becker (ASSL, 2009) p. 41
20. M.A. McLaughlin et al., Nature **439**, 817 (2006)
21. S. Mereghetti, The Astronomy & Astrophysics Review **15**, 225 (2008)
22. N. Rathnasree, J.M. Rankin, ApJ **452**, 814 (1995)
23. N. Rea et al., MNRAS (2010). astro-ph/1003.2085 (in press)
24. S.L. Shemar, A.G. Lyne, MNRAS **282**, 677 (1996)
25. C. Thompson, R.C. Duncan, ApJ **473**, 322 (1996)
26. N. Wang, R.N. Manchester, S. Johnston, MNRAS **377**, 1383 (2007)
27. P. Weltevrede, R.T. Edwards, B.W. Stappers, A&A **445**, 243 (2006)
28. W.Z. Zou, N. Wang, R.N. Manchester, J.O. Urama, G. Hobbs, Z.Y. Liu, J.P. Yuan, MNRAS **384**, 1063 (2008)

Chapter 6
Timing Solutions for Newly Discovered Sources

In the following chapter, Sect. 6.1 forms the basis of a paper, currently in preparation, although parts of Sect. 6.1.3 are from a paper published in the Monthly Notices of the Royal Astronomical Society, McLaughlin et al. 2009, *MNRAS*, **400**, pp. 1431–1438 (astro-ph/0908.3813).

In this chapter we present complete timing solutions for 7 of the newly discovered PMSingle sources reported in Chap. 4, all of which span at least 500 days. As well as reporting on the status of some provisional timing solutions for both PMSingle and other newly discovered sources. We discuss J1840−1419 which we identify as the most interesting source in our sample. We conclude with a recap of new timing solutions for the original RRATs and an outlook for future timing programmes.

6.1 PMSingle Timing Solutions

A followup campaign was performed for confirmation and timing of the sources discovered in the PMSingle analysis Sect. 4.4, in Parkes observing proposal P661 (P.I. Keane). Below we report the complete timing solutions for 7 PMSingle sources. We then give updates, and provisional timing solutions where appropriate, on the PMSingle sources which do not yet have a determined timing solution. This is followed with new discoveries resulting from analysis of a different Parkes survey (described below) and a discussion of the initial timing solutions for these sources.

6.1.1 Complete Timing Solutions

We have complete timing solutions, consisting of fits in period, period derivative, right ascension and declination, for 7 of the PMSingle sources. Figure 6.2 shows the timing residuals for 6 of these sources (all but J1652−4406) and Table 6.1 gives the parameters of the fits. Here we quickly review each of the sources in turn.

E. F. Keane, *The Transient Radio Sky*, Springer Theses,
DOI: 10.1007/978-3-642-19627-0_6,

Table 6.1 The timing solutions, both complete and incomplete, of the all PMPS RRAT sources

Source	RA (J2000)	DEC (J2000)	P (s)	$\dot{P}$ (10^{-15})	PEPOCH (MJD)	B (10^{12} G)	τ (Myr)	$\dot{E}$ (10^{31} erg s^{-1})
Complete PMSingle timing solutions								
J1513−5946	15:13:44.78(1)	−59:46:31.9(7)	1.046117156733(8)	8.5284(4)	54909	3.02	1.9	29.4
J1554−5209	15:54:27.15(2)	−52:09:38.3(4)	0.1252295584025(7)	2.29442(5)	54909	0.5	0.9	4605.9
J1652−4406	16:52:59.5(2)	−44:06:05(4)	7.707183007(4)	9.5(2)	54947	8.6	12.8	0.1
J1707−4417	17:07:41.41(3)	−44:17:19(1)	5.7637770030(4)	11.65(2)	54999	8.3	7.8	0.2
J1807−2557	18:07:13.66(1)	−25:57:20(5)	2.76419486975(4)	4.994(2)	54909	3.8	8.8	0.9
J1840−1419	18:40:32.96(1)	−14:19:05(1)	6.5975626227(4)	6.33(2)	54909	6.5	16.5	0.1
J1854+0306	18:54:02.98(3)	+03:06:14(1)	4.5578200962(1)	145.125(6)	54944	26.0	0.50	6.1
Original RRAT timing solutions								
J0847−4316	08:47:57.33(5)	−43.16:56.8(7)	5.9774927370(7)	119.94(2)	53816	25.1	0.8	2.0
J1317−5759	13:17:46.29(3)	−57:59:30.5(3)	2.64219851320(5)	12.560(3)	53911	6.3	3.2	2.5
J1444−6026	14:44:06.02(7)	−60:26:09.4(4)	4.7585755679(2)	18.542(8)	53893	10.0	4.0	0.6
J1819−1458	18:19:34.173(1)	−14:58:03.57(1)	4.26316403291(5)	575.171(1)	53351	50.1	0.1	32.8
J1826−1419	18:26:42.391(4)	−14:19:21.6(3)	0.770620171033(7)	8.7841(2)	54053	2.5	1.3	79.4
J1846−0257	18:46:15.49(4)	−02:58:36.0(2)	4.4767225398(1)	160.587(3)	53039	25.1	0.4	6.3
J1913+1330	19:13:17.975(8)	+13:30:32.8(1)	0.92339055858(2)	8.6799(2)	53987	2.5	1.6	39.8
Preliminary/unsolved PMSingle sources								
J1047−58	10:47(1)	−58:41(7)	1.23129(1)	–	55779	–	–	–
J1423−56	14:23(1)	−56:47(7)	1.42721(7)	–	54557	–	–	–
J1703−38	17:03(1)	−38:12(7)	6.443(1)	–	54999	–	–	–
J1724−35	17:24(1)	−35:49(7)	1.42199(2)	–	54776	–	–	–
J1727−29	17:27(1)	−29:59(7)	–	–	–	–	–	–
Unsolved original RRATs								
J1754−30	17:54(1)	−30:11(7)	1.26785(1)	–	53189	–	–	–
J1839−01	18:39(1)	−01:36(7)	0.93190(1)	–	51038	–	–	–
J1848−12	18:48(1)	−12:47(7)	6.7953(5)	–	53158	–	–	–
J1911+00	19:11(1)	+00:37(7)	6.94(1)	–	52318	–	–	–

J1513−5946

J1513−5946 (formerly J1514−59) maintains a 100% detection rate, from 30 observations, totalling 18 h. The periodic nulling described in Sect. 4.5.2 is detected in every observation. During the 'on' periods, J1513–5946 is detectable in a periodicity search. Figure 6.2 shows its timing residuals where we can clearly see two bands, symptomatic of two pulse components. Removing this banding, as for J1819−1458 (see Sect. 5.2.2), we obtain the timing solution with $\chi^2/n_{\rm free} = 4.2$. The fact that this is not equal to 1 is expected due to our fundamental violation of the stable profile assumption (see Sect. 5.1.1) and is due to the intrinsic variability in the single pulses, i.e. they are variable in both phase and pulse width. The 'on' times are not long enough, at ~1min, to be able to form stable profiles and result in less TOAs with lower error bars, but with the same scatter as shown in Fig. 6.2. The timing solution places J1513–59 amongst the normal pulsars in the $P - \dot{P}$ diagram (see Fig. 6.4), along with two of the original RRAT sources, with perhaps a slightly higher than average magnetic field strength.

J1554−5209

J1554–5209 (formerly J1554–52) is also detected in all observations, totalling 13 h. The timing residuals show three clear bands, which, upon removal, gives us a timing solution with $\chi^2/n_{\rm free} \sim 10$, which we again attribute to the intrinsic variability in the single pulses. In units of pulse periods it has by far the worst scatter in its residuals. J1554–5209 is also occasionally detectable in periodicity searches, although with less significance (i.e. $r > 1$ in all observations). It is noticeable in Fig. 6.4 as the outlying PMSingle source with the lowest period. Although it has the highest $\dot{E}$ in our sample we have not yet detected a signal in the Fermi data. It has a typical magnetic field for a pulsar and with $\tau = 0.9$ Myr it is the second 'youngest' PMSingle source.

J1652−4406: A Very Slow Pulsar

As discussed in Sect. 4.5.3, J1652−4406 was a Class 1 candidate identified in the PMSingle analysis. Despite showing 9 strong pulses in its discovery, confirmation was difficult. A small number of bursts have since been observed but none as strong as in the original survey observation. These borderline detections were not enough to conclusively confirm the candidate but it turned out J1652−4406 was sometimes detectable by folding. Looking for a folded signal was made possible by obtaining an initial period of $P = 7.70718$ s from period differencing of the discovery burst TOAs. Using this, and the dispersion measure at which the bursts peaked, as a starting point, each of 28 followup observations were folded and dedispersed into archives consisting of 1-min subintegrations. The surrounding $P - DM$ space was

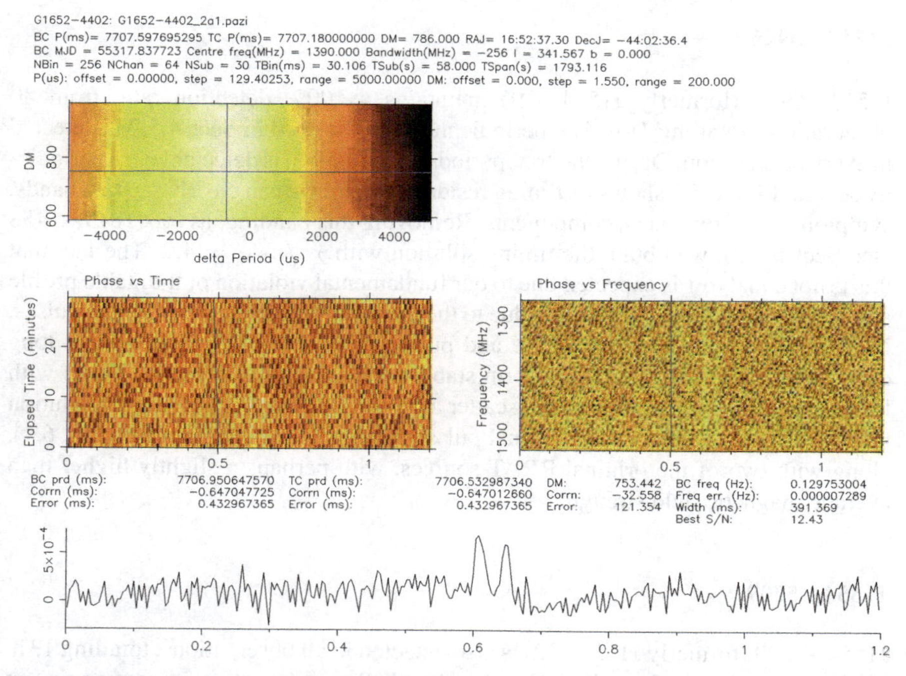

Fig. 6.1 The output of a search in $P - DM$ space about the initially determined values. (*Top*) The blue cross-hairs indicate the best values for this observation. (*Middle*) Sub-integrations (*left*) and dedispersed frequency channels (*right*) showing the best-fit line. The (*bottom*) panel shows the pulse profile where a clear double-peaked structure is seen

then searched for a folded signal using *PDMP*[1] (as shown in Fig. 6.1). In 22 of the observations a folded signal was detected with a double-peaked profile. From these a timing solution, spanning 560 days has been obtained with $\chi^2/n_{\text{free}} = 0.75$. This places J1652−4406 just above the death line, just like J1840−1419, but for this source there is little prospect of high energy followups as it is towards the Galactic centre, with $l = 341.570°$, $b = 0.0000°$ and with $DM = 786$ cm^{-3} pc has an inferred distance of 8.4 kpc [5]. At ten times further distance we expect 100 times less X-ray flux than from J1840−1419, but the situation is likely to be even worse given the extra absorption that would result from the large neutral hydrogen density in the Galactic centre. J1652−4406 is thus the third slowest radio pulsar known, just behind J1001−5939 ($P = 7.73$ s) and the famous slow pulsar J2144−6145 ($P = 8.51$ s).

J1707−4417

J1707−4417 (formerly J1707−44) has been detected in all but one of 23 observations which have totalled 13 h. The timing residuals show two clear bands, separated

[1] http://psrchive.sourceforge.net/manuals/pdmp

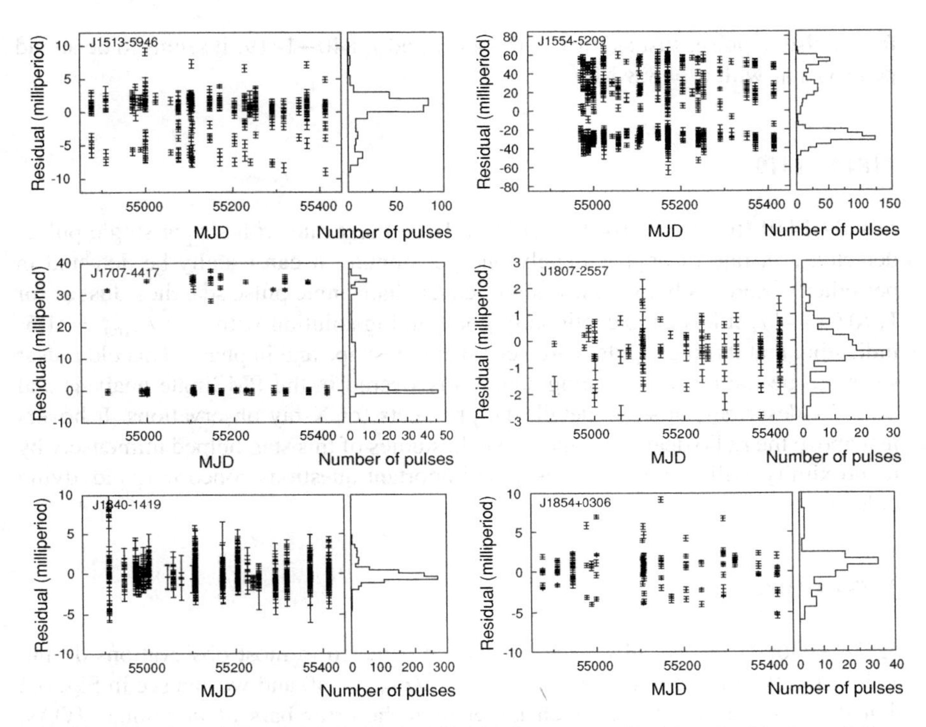

Fig. 6.2 Plotted are timing residuals for the 6 PMSingle sources with determined timing solutions. From top to bottom the sources are: J1513–5946, J1554–5209, J1707–4417, J1807–2557, J1840–1419, and J1854+0306. The ordinate in each plot is in units of milliperiods. Note the differing ranges for each source. This tells us that, for example, that J1840−1419 and J1707−4417 are much better single pulse 'timers' than J1554−5209

by ~200 ms. There are no other pulses detected between these two bands although there are instances where both pulse components are seen together. This suggests that the active emission time is longer than the time which the emission beam spends in our line of sight. Together these imply that J1707–4417 may have a 'patchy beam' [8]. Removing the banding effect, the timing solution we determine is remarkably good with $\chi^2/n_{\rm free} = 1.1$, indicating that the single pulses are very stable in phase and pulse width. J1707−4417 seems to be an old neutron star with $\tau = 7.8$ Myr and lies quite close to the death line, just above J1840−1419, in the $P - \dot{P}$ diagram.

J1807−2557

J1807−2557 (formerly J1807−25) is detected in all observations covering a total of 16h. The timing residuals do not show any obvious banding, although, in Fig. 6.2, there is a slight suggestion of a second band. The scatter in the residuals is quite large and the fit has $\chi^2/n_{\rm free} \sim 20$. Evidently the single pulses from this source are quite variable in phase. We can also see that the error bars in the TOAs vary quite a bit in extent, indicating that the shape of the individual pulses varies appreciably from pulse to pulse. Just as for J1707−4417 and J1840−1419, it seems to be an old neutron star with $\tau = 8.8$ Myr.

J1840−1419

J1840−1419 (formerly J1841−14) has a large burst rate with strong single pulses detected at a rate of approximately one per minute. It can usually be detected in periodicity searches but with less significance than single pulse searches. Just as for J1707−4417, it has an exceptionally good timing solution with a $\chi^2/n_{\rm free} = 1.5$, indicating that its single pulses are very stable in shape and in phase. This old pulsar seems to be the most interesting source discovered in the PMSingle analysis and Sect. 7.1 describes, in some detail, the prospects for X-ray observations. It hovers just above the radio death line and, as such, studies of this star, helped immensely by its proximity, will help us to investigate important questions concerning old, dying pulsars.

J1854+0306

J1854+0306 (formerly J1854+03) has been detected in most observations during 15 h of followup. The timing solution has $\chi^2/n_{\rm free} \sim 40$ and we can see in Fig. 6.2 that the observed scatter is much larger than the error bars of individual TOAs, indicating variability in pulse phase. The pulse widths are not seen to vary to the same degree. Of the PMSingle sources, J1854+0306 has the strongest and magnetic field, the second strongest of all the RRAT sources with determined B, behind J1819–1458 (see Table 6.1).

6.1.2 Preliminary/Unsolved PMSingle Sources

J1047−58 and J1423−56

These sources have been followed up as part of the P511 observing project (P.I. McLaughlin) at Parkes. No timing solutions had yet been determined at the time of writing.

J1703−38

J1703–38 is another source with a low burst rate of $\lesssim 2\,{\rm h}^{-1}$. Despite this, as we suggested in Sect. 4.5.2, we have been able to determine a period of $P = 6.443$ s using period differencing. A timing solution has not been forthcoming however as very long observations (>1h) are needed to guarantee the detection of multiple pulses (essential for identifying the topocentric period in each observation). Higher sensitivity observations and perhaps lower frequencies (where the burst rate might be higher) are planned for the future using the GBT.

J1724−35

J1724−35 was the first PMSingle candidate to be confirmed. It has been missed in 6 of 21 followup observations which have totalled 15 h. Furthermore, when detected its observed burst rate is $\lesssim 3\,\mathrm{h}^{-1}$, which is quite low, so that obtaining a coherent timing solution has not been possible. A renewed attempt will be made in the future to 'solve' this source, using higher sensitivity, again at lower frequencies. These criteria also suggest further observations might be best pursued with the GBT, and this is planned for future work.

J1727−29

J1727−29 has by far the lowest burst rate of any of our confirmed sources with just 4 pulses detected in 6 h. Further followup is pointless, as such a low rate makes determining a timing solution impossible. In fact we have not even determined the underlying period, if any, in this source. With pulses of $\sim$7 ms its maximum source size is constrained to be $\sim$2,100 km by causality. This is much larger than a neutron star but less than half the minimum radius for a relativistic white dwarf at the Chandrasekhar mass [9] so that we suspect a neutron star origin.

Single Observation RRATs

Of the 8 single observation sources described in Sect. 4.5.3, J1652−4406 has been confirmed and observed numerous times. The others remain unconfirmed, suggesting that they are sources with very low burst rates or, in some cases, single transient events. Re-observation attempts for these sources are ongoing.

6.1.3 Original RRAT Timing Solutions

Since the discovery of the original 11 sources, followup timing observations have been performed primarily at Parkes, as part of observing proposal P511, with a few additional observations undertaken at Arecibo. Numerous observations of J1819−1458 and J1913+1330 were performed at Jodrell Bank, as have been described in Chap. 5. In addition to the PMSingle sources, Table 6.1 lists the timing solutions which have been obtained for 7 of the 11 sources.

6.1.4 New Discoveries

Using the Parkes Telescope, two surveys have been performed off the Galactic plane, i.e. outside the region where the PMPS observations were. These were performed at intermediate and high Galactic latitudes of $5° < |b| < 30°$ [6, 7]. The surveys used the same specifications as the PMPS, except with faster time sampling of 125 μs and shorter pointings of 4.4 min. Recently, Burke-Spolaor and Bailes [2] have analysed these surveys in search of isolated bursts and presented 14 new sources, 7 of which were candidates which had never been confirmed. One of these unconfirmed sources was in fact re-detected by the authors soon after their publication (Burke-Spolaor, private communication), but 6 sources remained unconfirmed. As part of the P661 observing project, these 6 sources were observed in search of single pulses and 3 of these sources have been confirmed. Two of these sources have been regularly observed since January 2010 and both have provisional timing solutions, coherent over a timescale of $\gtrsim$200 days, at the time of writing.

J0735−62

J0735 − 62 is not detected in two followup observations, each of 10-min duration. Recently we have made a third, 30-min observation where it was easily detected, and thus confirmed for the first time, with 20 strong single pulses. Analysing the TOA differences, as described in Sect. 4.4, we determine a topocentric period of $P = 4.865(1)$ s, consistent with the initial estimate of $P = 4.862$ s period published by Burke-Spolaor and Bailes [2]. Additionally the two non-detections support their claim that the source suffers from severe scintillation. For this reason we do not yet know if this source is solvable using a reasonable amount of observing time, but a single lengthy observation is planned for P661 observing sessions in the October 2010–March 2011 semester, to investigate this very question.

J1226−32

The original detection of J1226–32 contained only 3 pulses but this was sufficient for Burke-Spolaor and Bailes [2] to predict a period of $P = 6.193$ s. We have confirmed this candidate and have observed 45 pulses in almost 3 h of followups, although in one third of the observations it is not detectable. We confirm the published period. Our provisional timing solution is coherent since January 2010 and regular observations as part of the P661 project are planned. These should reveal a full timing solution once the data span surpasses the one year mark.

J1654−23

The original detection of J1654–23 also consisted of just 3 pulses. We have confirmed this source and have determined a period of 545 ms, which differs from the published estimate of Burke-Spolaor and Bailes [2]. This is not very surprising given their small number of detected pulses. Interestingly, the period we determine, from 106 pulses detected in 2.3 h, is not at a different harmonic. This suggests that perhaps one of the 3 pulses initially identified was terrestrial in origin. As for J1226−32 we have a provisional timing solution, coherent since January 2010 and regular observations as part of the P661 project are planned. A full timing solution is expected for this source, once a year of monitoring has been made.

Unconfirmed Sources

In addition to the above three sources, we have attempted to confirm three other sources. We have observed J0923−31 and J1610−17 for 1.0 and 1.2 h respectively but have not been able to make a confirmation. We have detected 5-weak pulses from J1753−12 at the correct DM, during 1.3 h of observation, although we hope a more significant confirmation will come with time. We have not yet followed up these three sources for as long as the three new confirmations. This is, in some sense, by design, as these sources showed just 1, 1 and 3 pulses respectively in their discovery observations, so we decided to initially focus on the higher burst rate source (which were subsequently confirmed).

6.1.5 An Aside: The Perils of EFAC

If we have some data, and a model for that data, it is common to determine the $\chi^2/n_{\rm free}$ value for the model, to test its validity. Suppose the data are as shown in Fig. 6.3. In this example, if your model is a straight line of slope 1 you will get a large $\chi^2/n_{\rm free}$ value, suggestive of a 'bad fit'. This can be interpreted in two ways: (1) your model is incorrect; (2) your data is incorrect, for instance, you may have underestimated the errors in your data points. If you have $\chi^2/n_{\rm free} = {\rm EFAC}^2$ then scaling up your errors by a factor of EFAC will give you $\chi^2/n_{\rm free} = 1$, a good fit!

Such an operation is very bad practise, completely unjustified and should in no way be endorsed. Furthermore, forcing $\chi^2/n_{\rm free}$ to equal 1 completely defeats the purpose of the test for checking the validity of the model, i.e. you are simply accepting the model without testing it and setting the 'best fit' solution with this model to be the true solution. Unfortunately, the use of EFAC has actually been practised in pulsar timing studies in the past. We do not recommend this. For instance, if we had taken the $\chi^2/n_{\rm free}$ of our timing model for J1707 − 4417, ignoring the systematic signature of the two well-separated bands, we might apply an EFAC of 30. With an average measured TOA error of 2.8 ms, this means we would scale this up to ∼84 ms.

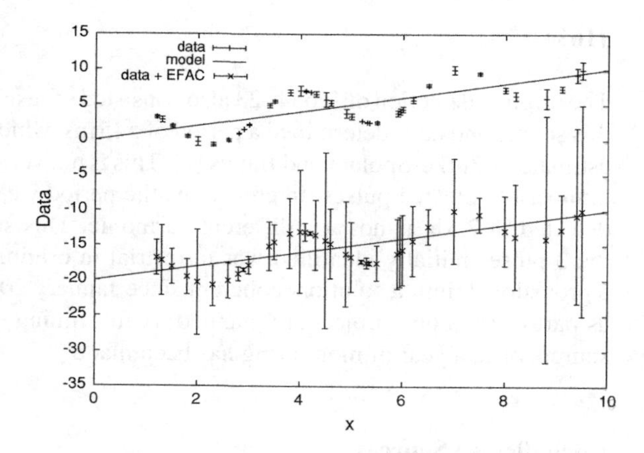

Fig. 6.3 This plot illustrates the dangers of using EFAC in pulsar timing. The $\chi^2/n_{\rm free}$ for the top straight-line fit to the data is large. If we blindly scale our errors to force $\chi^2/n_{\rm free} = 1$ then we can obtain a 'good fit' to a straight line but we have washed out the true underlying sinusoidal features

This essentially suggests that we can determine the peak of each pulse only to an accuracy of several times the pulse width (!) when in fact we expect a relationship of the form $\sigma_{\rm TOA} \approx W/(S/N)$. Even in the cases where all systematic trends have been removed there is still no justification for an EFAC, as our model cannot be perfect, e.g. the timing solutions presented above implicitly assume a stable pulse profile but this is clearly not the case for some of the sources with extra scatter in Fig. 6.2, above and beyond the size of the residuals. Another similar quantity sometimes used in pulsar timing is EQUAD, an error added in quadrature to all TOAs. However, just as there is no reason to assume all our measurements are incorrect by a constant factor, there is no reason to assume that all of our measurements are subject to an identical additive contribution from some unknown systematic. Use of EFAC and EQUAD is not recommended.

6.2 Timing Status and Prospects

We have reviewed the current status of timing solutions of RRATs. Of the 23 confirmed sources discovered in the PMPS, 14 now have coherent timing solutions. The prospects for obtaining solutions look bright for a couple of sources but seem unlikely for most, due, primarily, to the low rate of pulse detection, i.e. an unfeasibly long observing time would be required. A couple of the RRATs discovered in higher latitude Parkes surveys are expected to have full solutions in due course. This work on RRAT timing solutions is also an interesting exercise in single-pulse timing. Indeed, as we have shown, the procedure, starting at the bits recorded at the telescope, through to the complete timing solution, highlights many of the main features and assumptions of pulsar timing in general. We also note that these sources are the only pulsars with timing solutions obtained in this non-standard way. Figure 6.4 shows an up to date (late 2010) $P - \dot{P}$ diagram showing all known radio pulsars, with the 14 RRAT sources with known $\dot{P}$ identified, as well as the magnetars and XDINSs.

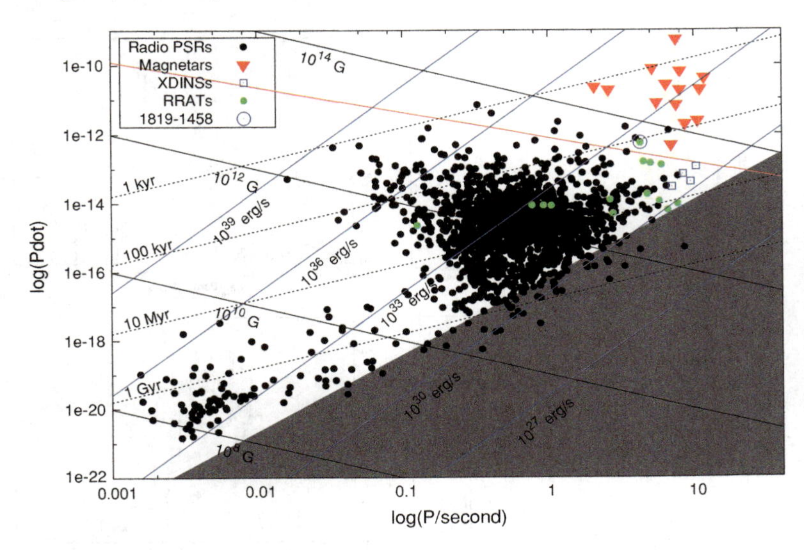

Fig. 6.4 The pulsar $P - \dot{P}$ diagram. Shown are the radio pulsars, which can clearly be seen to consist of two classes—the 'slow' pulsars and the MSPs, as well as those RRATs (J1819−1458 is *circled*), XDINSs and magnetars with known period derivative. The shaded region in the bottom right denotes the canonical 'death valley' of Chen and Ruderman [3] where we can see there is a distinct lack of sources. The radio loud–radio quiet boundary of Baring and Harding [1] is also shown (*orange line*) and we can see that only ~1% of sources are found above this line. Also plotted are lines of constant B, $\dot{E}$ and τ_c

6.2.1 Importance of Timing Solutions

The last 3 years has seen a large increase in our knowledge of the timing solutions of RRATs in the PMPS. Figure 2.5 summarised what was known in 2007: 11 known periods and 3 known period derivatives. Today, the figure is 22 known periods and 14 known period derivatives, obtained from observations at Jodrell Bank (see Chap. 5) and Parkes (see Sect. 6.1). Figure 6.5 shows (some of) the currently known radio properties of RRATs, and is an update of Fig. 6.1.

We can see from Figs. 6.4 and 6.5 that the RRATs certainly have long periods with half of the the 22 sources having periods $P > 4$ s. There are four sources with high inferred magnetic field strengths—they occupy a void region of $P - \dot{P}$ space and J1819−1458 remains the source with the highest B. At least four other RRATs (and possibly six) of the 14 with known $\dot{P}$ seem 'normal'. The remaining four sources have very long periods and are hovering over the death line. Identification of these 3 'groups' has only been possible through dedicated observing campaigns aimed at determining timing solutions, and hints at an answer (or rather answers) to the question: what is a RRAT? We discuss this in detail in Chap. 9 but it is clear that some are normal pulsars, some old/dying pulsars and some occupy the high-B void region of $P - \dot{P}$ space.

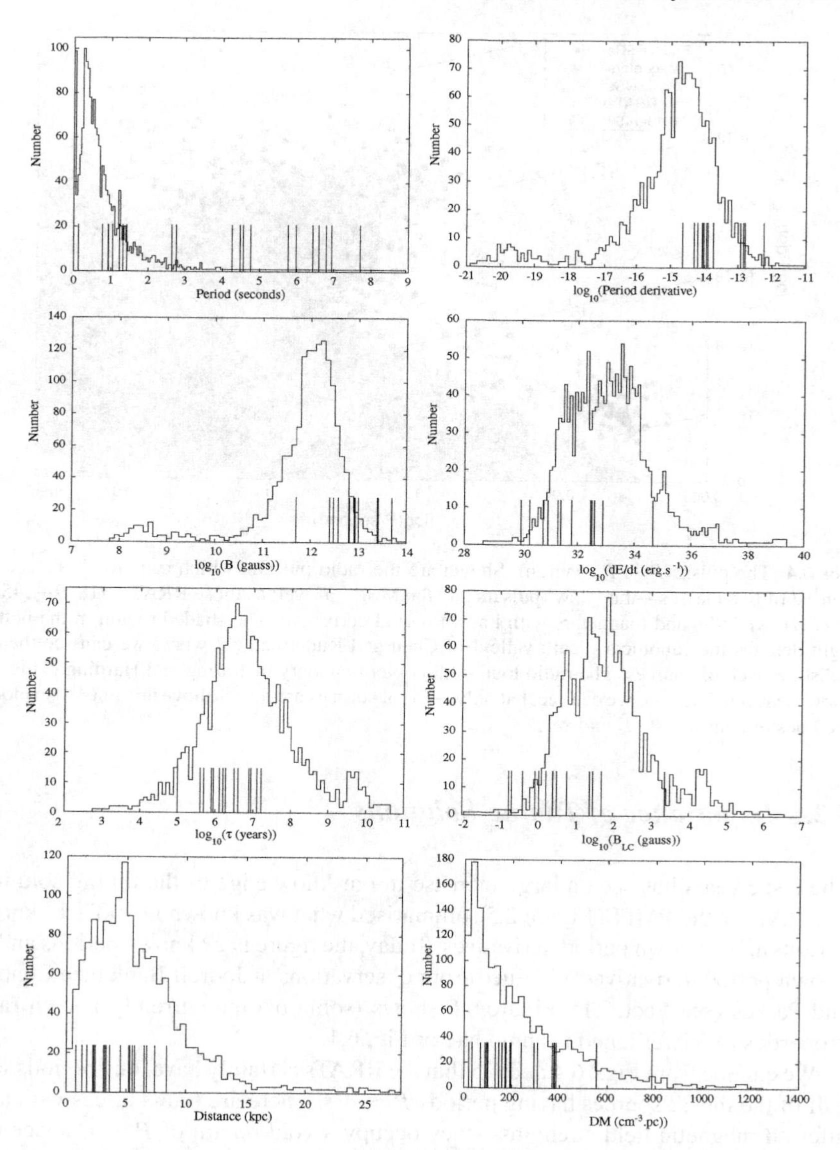

Fig. 6.5 In each plot the distribution is that of the overall radio pulsar population. Overplotted in each case are up to 23 impulses denoting the parameter values of the PMPS RRATs. The parameters plotted are: P, $\dot{P}$, B, $\dot{E}$, τ, $B_{\rm LC}$, distance and DM. In some cases the abscissa is plotted linearly, in some as a base-10 logarithm

Knowing where sources lie in $P - \dot{P}$ space allows us to investigate and/or infer pulsar evolutionary paths. For instance, we know that an isolated pulsar will move towards higher periods with time (recall pulsar current analysis described in Sect. 3.3), although the spin-down law is not known. With independent age estimates we can try to determine the path in $P - \dot{P}$ space of this evolution. Another intriguing possibility relates to the discovery of pulsars near the death line. The force-free magnetospheric models of Contopoulos and Spitkovsky [4] suggest a spin-down law of the form $n = 3 + 2/[(\Omega/\Omega_{\rm death}\cos^2\theta) - 1]$ which, at the death line, would be $n = 3 + 2\cos^2\theta/(1 - \cos^2\theta)$. This diverges for aligned rotators so that extremely high braking indices are implied at the death line, which would certainly be observable. Furthermore, more aligned sources are expected at the death line, owing to their slower energy loss-rate (see Eq. 2.35). The nearby pulsar J1840−1419 fits the bill for such investigations of old neutron stars as we will discuss in Sect. 7.1. Timing solutions give us estimates for a number of parameters (plotted in Fig. 6.5) such as B, $\dot{E}$, τ and $B_{\rm LC}$. Furthermore, we can identify additional contributions to pulsar spin evolution, in particular due to glitches. The timing solutions will also enable an improved retrospective search for pulses as we will be able to focus in on particular ranges of rotational phase. This will effectively allow an optimal nulling analyses, i.e. statistically we will be able to infer weaker pulses which may have been missed in the initial searches and this will help us determine what fraction of the time a source is truly 'off'. Timing solutions provide us with accurate astrometry, something which we would not have otherwise due to the poor spatial resolution of single-dish radio telescopes. This is crucial for observations at other wavelengths, which can be used to perform further investigations of the nature of the sources. For example, γ-ray observations probe the regime wherein the majority of the pulsar energy is output. We have reported a tentative detection of γ-ray emission from J1819−1458, although nothing is seen for the other sources (perhaps because, as we can see from Fig. 6.5, the RRAT $\dot{E}$ values are low). X-ray observations can provide tremendous insight and can tell us the spectra and temperature of neutron stars which enables the calibration of cooling curves and another form of age estimate. Additionally, any spectral lines may provide another estimate of B. Infrared observations enable investigation of surrounding fall-back discs and optical observations are useful in a search for GRP-related emission.

In the following chapter we describe a planned X-ray observation of J1840−1419 as well as simultaneous optical-radio observations of J1819–1458. Such observations would not be possible without coherent timing solutions.

References

1. M.G. Baring, A.K. Harding, ApJ **507**, L55 (1998)
2. S. Burke-Spolaor, M. Bailes, MNRAS **402**, 855 (2010)
3. K. Chen, M. Ruderman, ApJ **402**, 264 (1993)
4. I. Contopoulos, A. Spitkovsky, ApJ **643**, 1139 (2006)
5. J.M. Cordes, T.J.W. Lazio, preprint (arXiv:astro-ph/0207156)
6. R.T. Edwards, M. Bailes, van W. Straten, M.C. Britton, MNRAS **326**, 358 (2001)
7. B.A. Jacoby, M. Bailes, S.M. Ord, R.T. Edwards, S.R. Kulkarni, ApJ **699** (2009)
8. A.G. Lyne, R.N. Manchester, MNRAS **234**, 477 (1988)
9. S.L. Shapiro, S.A. Teukolsky, Black Holes, White Dwarfs and Neutron Stars. The Physics of Compact Objects. (Wiley–Interscience, New York, 1983)

Chapter 7
X-ray and Optical Observations of RRATs

In this chapter, Sect. 7.1 is based upon an accepted proposal for *Chandra* Cycle 12. The remainder of the chapter is an enhanced version of a paper, submitted to the Monthly Notices of the Royal Astronomical Society.

7.1 J1840–1419: The Coolest Neutron Star Ever Known?

Of all the sources identified in the PMSingle analysis, J1840–1419 may be the most important. We can see, from Fig. 7.1, that it occupies a position of $P - \dot{P}$ space between the high-magnetic field radio pulsars, the magnetars and the XDINSs and skirts the pulsar 'death line', where pulsar emission is thought to fail [2]. If RRATs are transitionary objects, in an evolutionary sense, linking normal radio pulsars, magnetars, XDINSs and 'dead' pulsars, then studies of J1840–1419 are of huge interest. With a dispersion measure of 19.4 cm^{-3} pc, its inferred distance is just 850 pc [4], closer than 97% of all pulsars. This makes J1840–1419 unique as a strong, nearby source in this crucial region of $P - \dot{P}$ space. As of August 2010 the positional uncertainty from our timing solution is ~1 arcsec, suitable for observations in the X-ray. As compared to J1819–1458, the only RRAT so far observed in the X-ray (but see Sect. 8.1 for a review of attempted observations of other sources), we note that it has a higher radio burst rate, and, crucially, J1840–1419 is four times closer to Earth. With this in mind we have prepared a proposal to observe this source with the *Chandra* X-ray Observatory during observing cycle 12 with an observation time of 30 ks (8.33 h), which has been accepted.

The *Chandra* observations will allow us to pursue a number of scientific goals.

(i) *Determine an X-ray spectrum.* We will measure a spectrum, and hence determine the temperature and any spectral features. The temperature determination will allow an independent constraint on the age. It will provide a useful data point for neutron star cooling curves at the coldest temperatures. As neutron stars as old as J1840–1419 are not typically in such close proximity, J1840–1419 may become the coldest neutron star with a measured temperature. The *XMM-Newton* observations of

E. F. Keane, *The Transient Radio Sky*, Springer Theses,
DOI: 10.1007/978-3-642-19627-0_7, © Springer-Verlag Berlin Heidelberg 2011

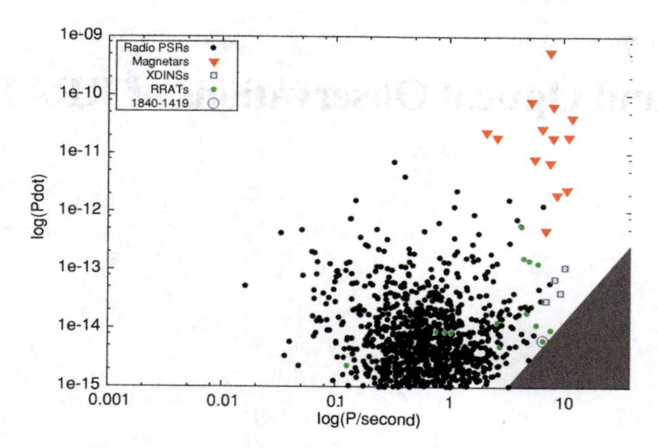

Fig. 7.1 The *top half* of the $P - \dot{P}$ diagram showing radio pulsars, RRATs, INSs, and magnetars, with lines of constant magnetic field and age. RRAT J1840–14 is located just on the pulsar "death line" [2]. Note that different models predict different locations for this line and it may in fact appear as a death 'band'. While it is assumed that pulsars are born in the upper left part of the diagram and move to the lower right as they age, the actual evolution of pulsars in this diagram and the relationships between different classes are unknown

J1819−1458detected unusual spectral features that may be cyclotron or atmospheric absorption lines. As we discussed in Sect. 2.3.2, these lines are rather unusual in radio pulsar X-ray spectra. A search for similar features in our J1840–1419 spectrum will reveal whether they are somehow related to the RRATs' unusual emission. Once again, the proximity of J1840–14 will help us, as we will be able to perform phase-resolved spectroscopy to search for any phase dependent variations in absorption line depth and temperature, as has been seen for at least two XDINSs [10].

(ii) *Search for periodicities*. We will search for X-ray pulsations at the radio-determined period. We will also be able to determine the X-ray pulsed fraction and absolute alignment with the radio bursts. This should allow us to determine whether the radio bursts originate in the polar cap or outer magnetosphere, i.e. if the radio pulses line up with the X-ray 'hot-spots' then it implies the polar cap is heated by the pairs associated with radio emission, a misalignment implies a different location for the radio emission.

(iii) *Search for X-ray/radio correlations*. We will simultaneously observe in the radio to search for any X-ray/radio correlations.

(iv) *Identify any surrounding pulsar wind nebula*. The observations will also have very high-resolution and so should be able to convincingly detect any surrounding features, such as the putative pulsar wind nebula claimed for J1819−1458, which we will discuss in Chap. 8.

This source also offers a unique opportunity to expand our knowledge of old neutron stars in general. Many young and middle-aged neutron stars have been studied at X-ray energies but few pulsars with ages greater than a Myr have been detected. This is unfortunate as these old pulsars appear to have surprising properties. The oldest detected neutron star, J0108–1431, has a characteristic age of 170 Myr (i.e. ten times older than J1840–1419). It was observed with *Chandra* for 30 ks and found to be surprisingly active, with an X-ray luminosity of roughly 0.4% of its spin-down luminosity and a high-blackbody temperature of 0.28 keV [19]. This is consistent with the suggestion by Harding and Muslimov [11, 12] that polar cap heating becomes more efficient as pulsars age. X-ray observations of a pulsar of very similar age to J1840–1419, B0950+08 ($\tau = 16$ Myr) show that the emission is composed of both power-law and thermal ($kT = 86$ eV) components (e.g. [28]). The energy-dependent pulse profile shape of this pulsar shows that the magnetospheric and thermal contributions vary with energy. Slightly older than B0950+08, J1840–1419 will in fact be the *second oldest* non-recycled pulsar ever detected in X-rays. Searching for similar phenomena in J1840–1419 will reveal whether these properties are common among old neutron stars.

Our 30 ks observation will utilise ACIS-S in the 1/8 sub-array mode with a 0.4 s frame rate and using the back-illuminated S3 CCD. This will result in 16.5 bins across the pulse profile, ample phase resolution for folding. For a neutron star with a radius of 10 km at a distance of 850 pc the unabsorbed bolometric flux is 1.5×10^{-11} erg s^{-1} cm^{-2} for an assumed $kT = 0.1$ keV (smaller than the $kT = 0.15$ keV determined for J1819–1458).Using COLDEN[1] to estimate the total neutral hydrogen density (to the edge of the Galaxy) in the direction of the source we get $N_{\mathrm{H,total}} = 4.3 \times 10^{21}$ cm^{-2}. A column density of $N_{\mathrm{H}} = 1.0 \times 10^{21}$ cm^{-2} (to the source) is appropriate as this source is nearby. For the ACIS-S setup this results in a count rate of 0.5 cps assuming 90% efficiency. In a 30 ks observation then we should obtain ~15,000 photons. This is roughly 12 times as many photons as were accumulated in the Rea et al. [22] *Chandra* observations (see Sect. 8.1) and 7 times as many photons accumulated as in the McLaughlin et al. [17] *XMM-Newton* observations of J1819–1458. This will allow a sensitive multi-component fit to the spectrum and an extremely sensitive search for any spectral absorption features. We will also be sensitive (at the 2σ level) to pulsed fractions as low as 5% for a duty cycle of 50% and to even lower pulsed fractions for smaller duty cycles.

If the temperature of RRAT J1840–1419 is more like the XDINSs with $kT \sim 80$ eV (see e.g. Haberl [10]) the unabsorbed bolometric flux is 6.1×10^{-12} erg s^{-1} cm^{-2} with a corresponding count rate of 0.14 cps (90% efficiency) resulting in 4,600 photons in 30 ks, still adequate for detailed spectral fits, a search for absorption features, and a 2σ sensitivity to pulsed fractions of 10% for a duty cycle of 50%. For the most pessimistic estimate, we take a temperature of just 50 eV which gives an unabsorbed flux of 9.3×10^{-13} erg s^{-1} cm^{-2}, a corresponding count rate of 6.142×10^{-3} cps (90% efficiency), and 165 counts in 30 ks. This count rate is still a factor of 300 above the background count rate and would enable a detection but no spectral fitting. We note that a temperature this low is extremely unlikely, given the higher temperatures for B0950+08, J1819–1458, and J0108–1431,and that much higher temperatures than mentioned in any scenario here are possible.

---

[1] http://cxc.harvard.edu/toolkit/colden.jsp

7.2 Optical Observations of J1819–1458

The reason why the radio emission from J1819–1458 is so highly variable is unknown. The possibilities that it is emitting giant radio pulses (see Sect. 2.3.2), or that it is somehow a transitionary object between the normal radio pulsars and magnetars (see e.g. Lyne et al. [15], or Sect. 5.2.2), both suggest that there may be detectable optical emission (see Shearer et al. [25] for observations of enhanced GRP-associated optical emission, as well as Stefanesen et al. [27] and Castro-Tirado et al. [1] for observations of a purported optically flaring magnetar). Additionally, with no a priori knowledge of the behaviour of its emission in different wavelength regimes, it makes sense to investigate, so that its spectral energy distribution can be characterised. In Sect. 2.3.2 we described X-ray observations of J1819–1458, and we will elaborate on these in Sect. 8.1. In Sect. 5.2.2 we reported the possible detection in γ-rays. J1819–1458 is regularly studied in the radio, at the opposite end of the electromagnetic spectrum. As we have described, the detectable radio emission from J1819–1458 amounts to only $\sim$1 s per day. Assuming that the optical light behaves in a similar manner, long exposures of the field would be relatively insensitive due to the accumulation of sky photons. A much better way of detecting optical emission from J1819–1458 would then be to observe with a high-speed optical camera simultaneously with radio observations, and co-add only those optical frames coincident with radio bursts. We present the results of such a search, using simultaneous ULTRACAM and Lovell Telescope observations.

Although deep infrared observations have revealed very tentative evidence for a counterpart at $K \sim 21$ magnitude ([21], see Sect. 8.1), there was, in the initial observations, no evidence for an optical counterpart, but this could be due to the rather modest magnitude limit of $I = 17.5$ [23]. Taking longer exposures to go deeper, however, is not necessarily the best solution, as the RRATs may have very faint persistent optical/IR emission and only emit strongly at these wavelengths during bursts. The main optical pulse of the Crab pulsar, for example, is $\sim$5 magnitudes brighter than its persistent light level [3, 9, 24]. In this case, the best strategy would be to reduce the contribution of the sky and take a continuous sequence of extremely short exposures on a large-aperture telescope covering a number of burst cycles in order to catch a burst in one or two of the frames. Dhillon et al. [5] have tried such an approach, using the high-speed CCD camera ULTRACAM [7] on the 4.2 m William Herschel Telescope (WHT) but found no evidence for bursts brighter than $i' = 16.6$. While this limit may not appear to be particularly deep, it must be remembered that it refers to the burst magnitude, not the persistent magnitude. In fact, there is only one way in which it is possible to significantly improve upon this [5] limit: observe in the optical simultaneously with the radio, which would allow just those optical frames coincident with the radio bursts to be searched for optical bursts. We performed such observations, obtained with ULTRACAM on the WHT and the 3.5 m New Technology Telescope (NTT) simultaneously with the 76 m Lovell Telescope at Jodrell Bank Observatory (JBO).

7.2.1 Observations and Data Reduction

The observations of J1819–1458 were obtained on the nights of 2008 August 6 (WHT+ ULTRACAM and JBO) and 2010 June 14 (NTT+ ULTRACAM and JBO). In addition to shot noise from any object flux, every ULTRACAM data frame has noise contributions from the sky and CCD readout noise. The sky noise can be reduced by reducing the exposure time, but the readout noise cannot. Hence it makes sense to expose each data frame for as long as the readout noise is the dominant noise source, thereby maximising the chances of observing a burst in a single frame without significantly degrading the signal-to-noise ratio of the data. ULTRACAM was hence used in drift mode, with one window centred on the X-ray position of the RRAT ([22], see Sect. 8.1) and the other on a nearby comparison star (see the top panel of Fig. 7.2). An SDSS i' filter and slow readout speed was used in the red arm of ULTRACAM on both nights, and the focal-plane mask was used to prevent light from bright stars and the sky from contaminating the windows (see [7] for details).

On 2008 August 6, the CCD windows were unbinned and of size 60×60 pixels, where each pixel on the WHT is $0.3''$. A total of 112,588 frames were obtained between 21:11 and 22:49 UTC on this night, each of 51.1 ms exposure time and 1.4 ms dead time. The data were obtained in photometric conditions, with no Moon and seeing of $0.9''$. On 2010 June 14, the CCD windows were binned 2×2 and of size 150×150 pixels, where each unbinned pixel on the NTT is $0.35''$. A total of 68,274 frames were obtained between 01:32–02:14 UTC and 02:45–03:46 UTC; the gap in the middle of the run was due to a GRB override observation (GCN circular 10,841, D'Elia et al.). Each frame had an exposure time of 86.5 ms and a dead time of 3.5 ms. Conditions on this night were not as good as in 2008, with seeing of $1.9''$ at the start of the run, dropping to $1.2''$ at the end. The night was photometric and there was no Moon.

Simultaneous radio observations at JBO were made at an observing frequency of 1.4 GHz as described in Sect. 5.1.3. The 2008 dataset utilised the AFB with bandwidth consisting of 64×1 MHz channels and a time sampling of 100 μs. The 2010 dataset was recorded using the DFB with a bandwidth of 1024×0.5 MHz channels and a time sampling of 1 ms. In both cases, the polarisations were summed to give total intensity (Stokes I) and the output was either 1-bit digitised (in 2008) or 2-bit digitised (in 2010).

The ULTRACAM frames were first debiased and then flat-fielded using images of the twilight sky. The list of JBO pulse arrival times on each night were then corrected to the TDB timescale at the solar system barycentre and compared with the barycentred ULTRACAM TDB times. Note that each ULTRACAM frame is time-stamped to a relative (i.e. frame-to-frame) accuracy of ~50 μs and an absolute accuracy of ~1 ms using a dedicated GPS system (see Dhillon et al. [7]). It was found that 24 and 25 ULTRACAM frames on 2008 August 6 and 2010 June 14, respectively, contained radio bursts, and these were then shifted to correct for telescope guiding errors and co-added (see Sect. 7.2.2).

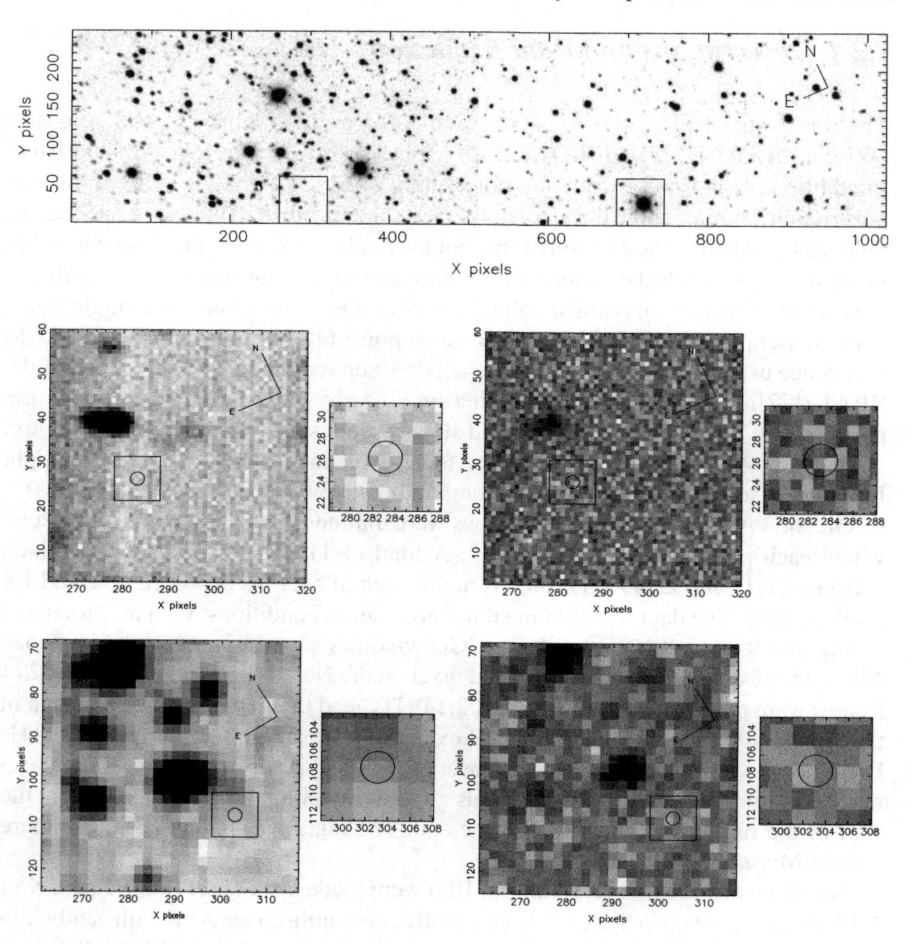

Fig. 7.2 (*Top*) The finder chart for the J1819−1458 region in the i' band, taken by summing 40 acquisition frames on 2008 August 6 with a total exposure time of 127 s. The boxes represent the 2 drift-mode windows used to acquire high-speed data on J1819−1458 (*left box*) and a bright comparison star (*right box*). The plate scale is $0.3''$/pixel and the orientation is denoted by the arrows. (*Middle*, *left*) The optical image obtained from adding all WHT frames. In the zoomed-in box, the $0.5''$-radius circle is centred on the X-ray position of J1819–1458. The plate scale is $0.3''$/pixel. (*Middle*, *right*) the co-added optical image using only those optical frames coincident with radio bursts. (*Bottom*, *left*) The NTT optical image, after addition of all frames. The plate scale in $0.7''$/pixel. (*Bottom*, *right*) the co-added NTT optical image using only the optical frames coincident with radio bursts

Aperture photometry at the X-ray position of J1819–1458 was also performed using the ULTRACAM pipeline data reduction system. To do this, we had to determine the pixel position of J1819–1458 on the ULTRACAM CCD. This was achieved by transforming the x, y pixel coordinates to equatorial coordinates using the known positions of bright stars in the field, using the second Guide Star Catalogue [14].

Incorporating the uncertainties in this transformation, the reference star positions and the X-ray position of J1819–1458, we estimate that the uncertainty in the resulting pixel position is 0.5″. We extracted a light curve for both the comparison star and the position of the RRAT using variable-sized apertures with radii set to 1.5 times the seeing, as measured from the FWHM of the comparison star, which is ~3–6 times larger than the error in the RRAT position on the ULTRACAM frames. The sky level was determined from an annulus surrounding each aperture and subtracted from the object counts.

7.2.2 Results

Radio Observations

The radio pulses detected from J1819–1458 are known to arrive preferentially at three distinct rotation phases (Lyne et al. [15], see Sect. 5.2.2). The top panel of Fig. 7.3 shows a grey-scale intensity plot of the individual pulses detected during the 2010 June 14 observation with the Lovell Telescope, as well as the combined profile from adding these pulses together. The three 'sub-pulses' are clearly visible. The bottom panel of Fig. 7.3 shows a histogram of pulse arrival times in rotation phase with respect to the long-term radio-derived ephemeris at JBO. This is essentially a probability distribution in rotation phase for the radio pulses. The unshaded histogram denotes all pulses detected, as part of our regular timing observations, in the time interval between the optical observations in 2008 and in 2010. The shaded (red) histogram shows the corresponding distribution for the pulses detecting during the optical observations. The two histograms are similar, implying that the pulses detected during the simultaneous observations were typical, and J1819–1458 seems to have been no more nor less 'active' in the radio than at other times.

Optical Observations

The sum of the 24 and 25 ULTRACAM i'-band frames containing radio bursts on 2008 August 6 and 2010 June 14 are shown in the right-hand central and lower panels of Fig. 7.2. For comparison, the sum of all the ULTRACAM frames obtained on each night are shown in the corresponding left-hand panels.

The circles plotted in Fig. 7.2 indicate the expected position of J1819–1458, with the radius equal to the error in this position on our ULTRACAM frames. Inspecting the zoomed-in boxes to the right of each panel reveals no visual evidence for the RRAT in either burst (right) or persistent (left) light. Note the significantly worse quality of the NTT data (bottom panels in Fig. 7.2) compared to the WHT data (central panels) due to the poorer seeing, which forced us to on-chip bin these data by a factor of 2.

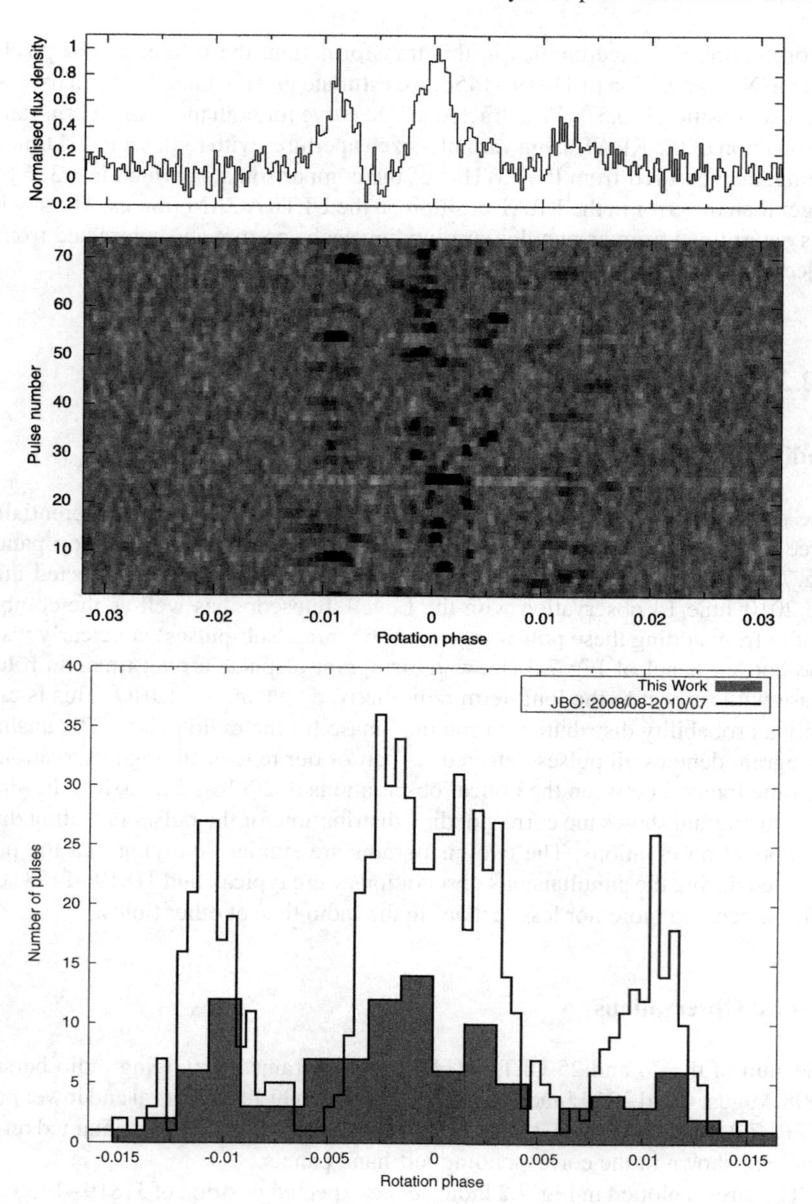

Fig. 7.3 (*Top*) Grey plot showing a 'pulse-stack' of radio pulses detected from J1819−1458during the Lovell-NTT observation, and on *top*, the integrated pulse profile showing the three characteristic peaks (see e.g. Lyne et al. [15] and Karastergiou et al. [13]). (*Bottom*) Histogram of radio bursts from J1819−1458, during the observations reported here, and for all bursts detected at Jodrell Bank in the past two years

So far, we have implicitly assumed that the optical and radio bursts are coincident, and that the radio bursts are equal or shorter in duration than the putative optical bursts. This is a reasonable assumption given, for example, the behaviour of the Crab pulsar, which shows that the optical pulse is approximately five times wider than the radio pulse and leads the radio pulse by $\sim$200 μs [18, 25, 26]. Since we know precisely when the radio bursts occurred, it is a simple matter to search for optical bursts lagging/leading the radio bursts and/or of different widths to the radio bursts by combining the appropriate optical frames. Hence, as well as summing the ULTRACAM frames coincident with the radio bursts, n, we also co-added the frames $n-1, n+1$ and $n \pm 1$. None of the resulting images show any evidence of the RRAT.

The searches described so far have relied on the ability of the eye to identify a star in a summed image. A different approach is to inspect the light curve obtained by extracting the counts in an aperture centred on the position of the RRAT. Figure 7.4 shows the light curves obtained on each night. The signature of an optical counterpart to the radio bursts would be a series of deviant points lying approximately 5σ or greater from the mean[2] and aligned temporally with the radio bursts (indicated by the vertical tick marks near the top of Fig. 7.4). Given that the exposure time is significantly longer than the radio burst duration, one would expect only one, or at most two, points per burst.

The dashed line in Fig. 7.4. shows the $+5\sigma$ deviation level. It can be seen that there are no points lying above this line. Moreover, the optical points coincident with the radio bursts, marked by the red circles in Fig. 7.4, appear to be randomly scattered about the mean level of zero. The implication is that we have not detected any evidence for optical counterparts to the radio bursts from J1819−1458. We also searched for periodicities in the light curves using a Lomb-Scargle periodogram [20]. No evidence for a significant peak around the 4.263 s rotation period, or any other period, was found. As stated above, the dead-time of ULTRACAM during our J1819−1458 observations was always an insignificant fraction of the exposure time and approximately the same duration as the radio bursts shown in Fig. 7.3. This makes it unlikely we missed a single optical burst whilst ULTRACAM was reading out, let alone the expected 24–25 bursts.

It is useful to place a magnitude limit on the optical bursts from J1819−1458 in order to constrain the spectral energy distribution. From the summed images of the burst frames, we find that the RRAT shows no evidence for optical bursts brighter than $i' = 19.3$ at the 5σ level. As expected, the simultaneous radio observations have enabled us to impose a significantly deeper limit than reported in Dhillon et al. [5], who derived $i' > 16.6$. The corresponding flux density limit is $i' < 70\,\mu$Jy, where the flux has been calculated using Eq. 2 of Fukugita et al. [8], namely $f_\nu(\mathrm{Jy}) = 3631 \times 10^{-0.4i'}$, and the effective wavelength of the observation is 7610 Å (see Dhillon et al. [5]). We can now compare this to the (pulsed) radio flux density of 3,600 mJy at 1.4 GHz measured by McLaughlin et al. [17] to deduce that

[2] We have chosen 5σ as we have $\sim 10^5$ data points and only one point in $\sim 10^6$ would be expected to be greater than 5σ from the mean in a Gaussian distribution.

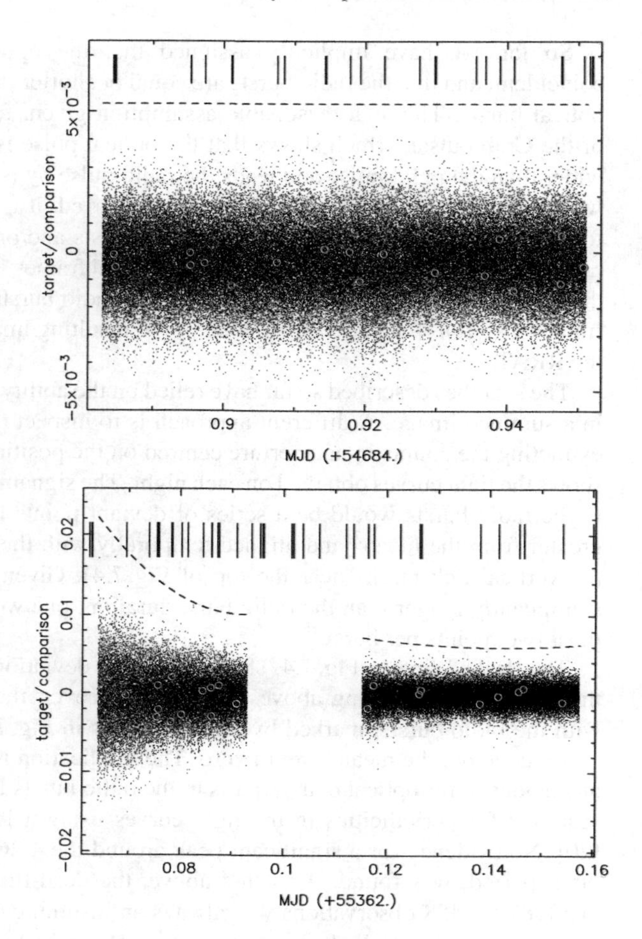

Fig. 7.4 The optical i'-band light curves of the J1819−1458region for the 112,558 WHT frames (*top*) and the 68,274 NTT frames (*bottom*). The *dashed lines* denote the 5-σ threshold, where we would expect 1 noise event during the observations. The *red circles* denote the optical frames coincident with radio pulses as detected by the Lovell Telescope at Jodrell Bank. The increased scatter at the start of the NTT observation was due to poor seeing. To plot the *dashed* $+5\sigma$ curve in this case, the standard deviation was calculated for groups of 100 points and the result fitted with a polynomial.The gap in the centre of the run was due to a GRB override observation with ULTRACAM

the spectral slope must be steeper than approximately $f_\nu \propto \nu^{-0.9}$. For comparison, the radio-to-optical slope of the *pulsed* radiation from the Crab has a much shallower slope of ~ -0.2 (measured from Fig. 9.3 of Lyne and Smith [16]), suggestive of different emission mechanisms at work. Note that the *unpulsed* X-ray emission of J1819−1458 (10 nJy at 0.3–5 keV [22]) also lies very close to the line $f_\nu \propto \nu^{-0.9}$.

7.2.3 Discussion

We find no evidence for optical bursts from J1819−1458 to a 5σ limit of $i' = 19.3$, 3 magnitudes deeper than the initial attempt by Dhillon et al. [5]. This allows us to say that the slope of the pulsed radio-optical spectrum must be steeper than −0.9, much steeper than for the pulsed emission from the Crab, and that extrapolating this slope correctly predicts the X-ray flux. There is thus no evidence for GRP-associated optical bursts.

We also see no evidence for any magnetar-like flares, although in comparison with the AXP 1E 1,048.1–5,937, for example, which has a magnitude of $i' = 25.3$ [6], our limit on J1819−1458 does not appear to be particularly deep. To place this in some context, note that if we had taken a single 1 h exposure of the field with the WHT under identical conditions, and assuming the object emitted 24 bursts, each of $i' = 19.3$ and 51.1 ms duration, we would have detected the object at only $\sim 0.02\sigma$. Using the high-speed photometry technique described in this paper, on the other hand, we would have detected the source at 5σ. The difference in sensitivity between the two techniques is due to the fact that the long exposure would be sky limited, whereas our approach is readout-noise limited. The only way we could now significantly improve upon our magnitude limit is to observe simultaneously for a longer period of time (in order to detect and co-add more bursts) and/or use a larger aperture telescope (in order to increase the number of counts detected from each burst). The discussion above assumes, of course, that the optical and radio light behave in a similar manner. If, however, the optical light has only a low (or no) pulsed fraction, then deep, long-exposure imaging might prove fruitful for optical identification, as might deeper searches for pulsed light on the proposed rotation period of the neutron star (e.g. Dhillon et al. [6]).

References

1. A.J. Castro-Tirado et al., Nature **455**, 506 (2008)
2. K. Chen, M. Ruderman, ApJ **402**, 264 (1993)
3. W.J. Cocke, M.J. Disney, D.J. Taylor, Nature **221**, 525 (1969)
4. J.M. Cordes, T.J.W. Lazio, preprint. (arXiv:astro-ph/0207156)
5. V.S. Dhillon, T.R. Marsh, S.P. Littlefair, MNRAS **372**, 209 (2006)
6. V.S. Dhillon et al., MNRAS **394**, L112 (2009) (astro-ph/0901.1559)
7. V.S. Dhillon et al., MNRAS **378**, 825 (2007)
8. M. Fukugita, T. Ichikawa, J.E. Gunn, M. Doi, K. Shimasaku, D.P. Schneider, AJ **111**, 1748 (1996)
9. A. Golden, A. Shearer, G.M. Beskin, ApJ **535**, 373 (2000)
10. F. Haberl, Astrophys. Space Sci. **308**, 171 (2007) (astro-ph/0609066)
11. A.K. Harding, A.G. Muslimov, ApJ **556**, 987 (2001)
12. A.K. Harding, A.G. Muslimov, ApJ **568**, 862 (2002)
13. A. Karastergiou, A.W. Hotan, W. van Straten, M.A. McLaughlin, S.M. Ord, MNRAS **396**, 95 (2009) (astro-ph/0905.1250)
14. B.M. Lasker et al., A J **136**, 735 (2008)
15. A.G. Lyne, M.A. McLaughlin, E.F. Keane, M. Kramer, C.M. Espinoza, B.W. Stapper, N.T. Palliyaguru, J. Miller, MNRAS **400**, 1439 (2009)
16. A.G. Lyne, F.G. Smith, Pulsar Astronomy. 3rd edn. (Cambridge University Press, Cambridge, 2004)
17. M.A. McLaughlin et al., Nature **439**, 817 (2006)
18. T. Oosterbroek et al., A&A **488**, 271 (2008)
19. G.G. Pavlov, O. Kargaltsev, J.A. Wong, G.P. Garmire, ApJ **691**, 458 (2009)
20. W.H. Press, G.B. Rybicki, Ap J **338**, 277 (1989)

21. N. Rea et al., MNRAS (2010), in press. (astro-ph/1003.2085)
22. N. Rea et al., ApJ **703**, L41 (2009)
23. S. Reynolds et al., ApJ **639**, L71 (2006)
24. A. Shearer, in *Astrophysics and Space Science Library*, vol. 351, ed. by D. Phelan, O. Ryan, A. Shearer (ASSL, 2008), p. 1
25. A. Shearer, B. Stappers, P. O'Connor, A. Golden, R. Strom, M. Redfern, O. Ryan, Science **301**, 493 (2003)
26. A. Słowikowska, G. Kanbach, M. Kramer, A. Stefanescu, MNRAS **397**, 103 (2009)
27. A. Stefanescu, G. Kanbach, A. Słowikowska, J. Greiner, S. McBreen, G. Sala, Nature **455**, 503 (2008)
28. V.E. Zavlin, G.G. Pavlov, ApJ **616**, 452 (2004)

Chapter 8
RRATs: An Overview

In the following chapter we review the work of many other authors in the field, during the past few years.

8.1 Recent Observations

Radio Observations

Much followup observations of the original 11 RRATs have been performed by McLaughlin and colleagues, using the Green Bank (GBT) and Arecibo Telescopes. Although not yet widely published, some of the results of this work are summarised in McLaughlin [25]. Observations at 350 MHz at the GBT have shown increased pulse rates for J0847−4316, J1754−30 and J1848−12. In the case of the first two sources, the rates were sufficiently high that FFT detections were made for the first time. However, the opposite is seen for J1819−1458, J1839−01 and J1846−0257, which were not detected at all, and J1826−1419 shows just one detected pulse. In observations at 327 MHz with Arecibo both J1911+00 and J1913+1330 are reported to have higher pulse rates. J1913+1330 is also reported to show clustering of pulses, which, as we have mentioned in Sect. 5.2.3, evident in the Jodrell Bank observations at L-band also. The remaining two sources, J1317−5759 and J1444−6026 have not been observed at either GBT or Arecibo due to their southerly positions.

Recently, all the Parkes observations of the original sources have been re-processed, focusing on the 8 sources with the highest burst rates (the 7 sources with timing solutions and J1754−30). Palliyaguru et al. (in preparation) have examined the distributions of pulse arrival times over timescales of months–years. They conclude that the distributions are not random, i.e. they are inconsistent with a uniform probability distribution, for 6 of the 8 sources. They have also searched for quasi-periodic behaviour but find no high significance signals. Miller et al. (submitted) have determined pulse amplitude distributions for these same 8 sources.

E. F. Keane, *The Transient Radio Sky*, Springer Theses,
DOI: 10.1007/978-3-642-19627-0_8, © Springer-Verlag Berlin Heidelberg 2011

They find, as seen for the initial PMSingle analysis in Sect. 4.6.2, that the distributions are well described as log-normal. Interestingly, for J1754−30, they report that the distribution does not turn over before hitting the noise floor, implying that it may have more underlying emission detectable with increased sensitivity (consistent with what is seen in the GBT observations of McLaughlin [25]. For J1819−1458 they conclude the opposite—increased sensitivity does not seem to result in the detection of more pulses so that J1819−1458 seems to be truly nulling. The spectral index measurements of Miller et al. are less conclusive but suggest that RRATs have spectra consistent with, or perhaps steeper, than typical radio pulsars and a flat radio-magnetar-like spectrum seems to be ruled out. It is unclear what these average spectra tell us given that the instantaneous values may vary hugely—indeed Miller et al. suggest just this in claiming that the emission from J1819−1458 is to some degree narrowband.

The only polarisation study[1] which has been carried out has been that of Karastergiou et al. [19] who have studied J1819−1458. As they have pointed out, there is a dearth of polarisation studies for high-B pulsars except for the 3 radio magnetars which show $\lesssim$100% linear polarisation. As typical for pulsars with similar, average $\dot{E}$ values, J1819−1458 shows 20–30% linear polarisation in its average profile, although the polarisation of individual pulses ranges from 0−100%, and very little circular polarisation. Karastergiou et al. also determined the rotation measure of J1819−1458 to be $\approx 330 \pm 30\,\mathrm{rad\,m^{-2}}$.

Latest Discoveries

In addition to the 30 RRATs which have been identified in the PMPS, several groups have performed searches in other pulsar surveys.

Arecibo RRATs: The PALFA survey, using the Arecibo telescope has observed the regions $30° < l < 78°$ and $162° < l < 214°$ at $|b| < 5°$. Like the PMPS, this survey was performed at 1.4 GHz, but with a bandwidth of 100 MHz spanned by 256 channels and with 64 μs sampling. Recently Deneva et al. [11] announced the discovery of 7 sources through single pulse searches, one of which, as we have previously noted is the PMSingle source J1854+0306.

Green Bank RRATs: A pulsar survey at 350 MHz using the GBT has been performed focusing on the region $75° < l < 165°$ and $|b| < 5.5°$, a region impossible to observe with the Arecibo or Parkes telescopes. The survey utilised a bandwidth of 50 MHz divided into 1,024 channels with 81.92 μs sampling. Initial results, processing the survey at a reduced resolution, have reported the discovery of 33 sources, 1 of which is a RRAT [16]. Another survey was performed with the GBT during the summer of 2007. A drift-scan was performed whilst the dish was immobilised due

[1] We have attempted to perform polarisation studies on the PMSingle sources at Parkes as a piggyback project to the PMSingle timing programme. However, the observing system suffered a number of both hardware and software problems meaning that we were unable to obtain any useful data. As of late, these issues seem to have be resolved (W. van Straten, private communication).

to track refurbishment. Although the results of the survey are as yet unpublished it has discovered 25 new sources, one of which is a RRAT [5].

Parkes RRATs: Burke-Spolaor and Bailes [6] have identified 14 new sources in single-pulse searches of two high-latitude pulsar surveys at Parkes (as we have described in Sect. 6.1.4). Of these, 8 had been observed multiple times but 6 were unconfirmed. We have confirmed 3 of these and are timing 2 of them.

Puschino RRATs: Recently Shitov et al. [35] have reported the discovery of a source showing 2 strong bursts in 3 hours of observation as well as sequences of weaker pulses lasting for tens of seconds.

Westerbork RRATs: The 8gr8 survey used the Westerbork Synthesis Radio Telescope array to survey the Cygnus region, which is known to harbour a number of OB associations [17]. The survey was performed at 328 MHz with a bandwidth of 10 MHz divided into 512 channels and using a time sampling of 819.2 μs. The detection of 4 RRAT sources has recently been reported [34] although one of these sources has a completely undetermined position.

Considering all of these surveys the total number of radio transient sources discovered as RRATs, at the time of writing, amounts to $30+6+1+14+1+4=56$.

X-ray Observations

Based solely on spin-down properties we have seen that RRATs, XDINSs and magnetars are similar. Thus it makes sense to observe RRATs in the X-ray to investigate how similar or not they may be in this regard. Perhaps, for instance, XDINSs are simply RRATs with unfavourable beaming. Of course, we also want to search for any magnetar-like X-ray flares.

We have already described the X-ray observations of J1819−1458 (see Sect. 2.3.2), made serendipitously by Renolds et al. [33] using *Chandra* (30 ks) and later by McLaughlin et al. [26] using *XMM-Newton* (43 ks). Further X-ray observations have been made by Rea et al. [32] using *Chandra* (30 ks). These authors report evidence for extended emission up to 5.5 arcsec from the star. Rea et al. suggest that the emission is due to a pulsar wind nebula (PWN) powered by the wind from J1819−1458. However this implies a rather large X-ray efficiency of $L_{\mathrm{X},0.3-5.0\,\mathrm{keV}}/\dot{E} \approx 0.2$, compared to typical values of $10^{-6}-10^{-1}$ [13] would see J1819−1458 as the lowest $\dot{E}$ source with such a PWN [20], and, unusually, may be required to be "magnetically powered" in some unspecified way in order to make sense. Rea et al. conclude that they cannot give an explanation for the extended emission.

In addition to J1819−1458, the two other RRATs with determined timing solutions in 2007, and hence those with sufficiently accurate astrometry, were observed in the X-ray. J1317−5759 was observed with *XMM-Newton* (32 ks) but with no detection [25, 31]. An unabsorbed 0.3–5.0 keV luminosity limit of 8×10^{32} erg s^{-1} was obtained,[2] lower than the X-ray luminosities of the XDINSs. No pulsations

[2] In comparison, J1819−1458 has a value of 4×10^{33} erg s^{-1} [26].

were detected. As we mentioned in Sect. 5.2.3, J1913+1330 was also observed, using *Swift* (9.3 ks), but with an incorrect source position [25, 30, 31]. When improved astrometry became available for J0847−4316 and J1846−0257 [25], these were also observed with *Chandra*, but no detections were made [18]. Luminosity and temperature limits of 1×10^{32} erg s^{-1}, 77 eV and 3×10^{32} erg s^{-1}, 140 eV were obtained, for J0847−4316 and J1846−0257 respectively. The observations of J0847−4316 ($D = 3.4$ kpc) consisted of just a single 10.7 ks pointing but a total of 196.1 ks is available for J1846−0257. This is due to a number of observations of the nearby X-ray pulsar J1846−0258, in the supernova remnant Kesteven 75, which shows no radio emission [2]. Although the distance to this source is uncertain ($D = 5.1 - 7.5$ kpc) if both sources were at 5 kpc then their angular separation of 2.5 arcmin implies they are a mere 3.6 pc apart. Kaplan et al. [18] have speculated that these stars may have once been in a binary, with the RRAT ejected upon supernova, and the X-ray pulsar formed in a second supernova, whose remnant is what is now visible as Kes 75. This is interesting as J1846−0258 has shown magnetar-like X-ray bursts [14], and as magnetars are thought to be formed from more massive progenitors [12, 28], although see Davies et al. [10], it suggests that the RRAT, which has $B = 2.5 \times 10^{13}$ G, having evolved first, may in fact be a magnetar progenitor.

Looking through the *Chandra*[3] and *XMM-Newton*[4] archives tells us that there has been a 97 ks observation of J1819−1458 using *XMM-Newton*, with simultaneous radio observations. This is as yet unreported but a publication is in preparation (M. A. McLaughlin, private communication). It will also be observed for 90 ks with *Chandra* in April 2011. Other than this the only other planned X-ray observation, at the time of writing, is of J1840−1419, which will also take place in 2011, with a duration of 30 ks.

Optical and Infrared Observations

Recently, Mignani [27] has reviewed the status of optical observations of neutron stars. Besides the observations described in Chap. 7, there have been no attempted optical observations of RRATs. On the other hand, infrared (IR) observations have recently been reported by Rea et al. [30]. Again, they observed the 3 RRATs which were first to have coherent timing solutions, using the near-IR camera NACO mounted on the Very Large Telescope in Chile. For J1317−5759 they find no IR counterpart down to a limiting magnitude of $K_s \sim 21$ and for J1913+1330 the observations again used an incorrect source position, so that this source lay outside of their field of view. For J1819−1458 they report a putative counterpart with $K_s = 20.96 \pm 0.10$ just on the edge of the $1 - \sigma$ error circle of the accurate X-ray position. Rea et al. conclude the identification of this counterpart with J1819−1458 needs to be further confirmed due to the high probability of a chance alignment.

[3] http://cda.harvard.edu/chaser/

[4] http://xmm.esac.esa.int/xsa/

γ-ray Observations

To our knowledge, there have been no reported γ-ray observations of RRATs. Following on from our possible detection of J1819$-$1458 (see Sect. 5.2.2) in data from the Fermi satellite we plan to search for γ-ray emission from all RRATs with coherent timing solutions.

8.2 Other Relevant Work and Ideas

In Sect. 5.1.1, we discussed the variable behaviour of pulsars on a period-by-period basis, e.g. the sub-pulse drifting shown in Fig. 5.1. Of relevance to the discussion of RRATs is nulling—the particular type of mode-changing where one mode shows a complete lack of radio emission. As typically observed, nulling occurs for 1–10 rotation periods, but the observed selection of nulling pulsars is quite biased [36]. Pulsars with longer nulling fraction are less likely to be detected in a single survey pointing, and in a confirmation pointing, and hence may be discarded amongst the plethora of pulsar candidates produced in modern surveys. Also, due to a lack of sufficient signal-to-noise ratio, weaker pulsars cannot be examined on shorter timescales. Thus there may well be nulling occurring either unnoticed or undetectable in many known pulsars.

As well as the RRATs, 2006 also saw the discovery of 'intermittent pulsars', sources which behave as normal radio pulsars for several days before switching off entirely for days to weeks. This switching occurs in a quasi-periodic fashion with PSR B1931+24, the archetypal system, turning 'on' for 5–10 days and 'off' for 25–35 days [21]. These timescales allow the measurement of separate slow-down rates during the on and off states, $\dot{\nu}_{\rm on}$ and $\dot{\nu}_{\rm off}$. The difference in these rates is about 50% which seems to reflect the extra energy loss due to the pulsar wind when there is radio emission, i.e. when off, the star slows down via dipole braking alone (the vacuum scenario of Chap. 2), and the magnetospheric plasma density decreases hugely. The on–off transition has been observed and lasts less than 10 s, indicating a massive change in magnetospheric currents on a very short timescale to a new state which is apparently stable for $\sim 10^6$ periods before switching once more. The explanation as to why this switching is quasi-periodic is unknown. Recently Lyne et al. [23] have shown switching between two stable states in 17 pulsars. Like in the case of PSR B1931+24, switches between two slow-down rates are observed with changes ranging from 0.3 to 13.3%, with highly-correlated pulse shape changes, i.e. moding. These results further strengthen the claim of Wang et al. [36] and others that moding and nulling are due to the same underlying phenomenon. Also, the data are not inconsistent with such switching being a generic property of all pulsars (A. G. Lyne, private communication).

Also of relevance to transient neutron stars is the pulsar 'death valley', where radio emission is thought to fail. Pulsar emission requires a supply of particles from the stellar surface which can be accelerated in the pulsar magnetosphere for pair production, $\gamma \rightarrow e^+ + e^-$, to ultimately lead to coherent radio emission. The strength of the electric potential ΔV depends on B and P, e.g. in the toy model of Goldreich and Julian [15] $\Delta V \propto B/P^2$. An electron accelerated in this potential will acquire a Lorentz factor of $e\Delta V/m_e c^2$. Depending on the emission mechanism (i.e. the dependence of ΔV on B and P) the minimum Lorentz factor sufficient for pair-production (the photon must have energy of at least $2m_e c^2$) defines a 'death-line', separating regions of $P - B$ space where radio pulsar emission is possible and regions where it is inhibited (the 'death valley'). Detailed considerations lead to different death-lines for different field configurations, e.g. on high curvature field lines [9], and death-lines for several emission mechanisms have been proposed (see e.g. [3, 29, 40]). However, the various death-lines do not satisfactorily explain the observed pulsar population and there is at least one pulsar which flouts the rules in the death valley, namely the 8.5 s PSR J2144−3933 whose detection as a radio pulsar poses serious challenges to emission theories [39].

The magnetars are thought to be isolated neutron stars with very strong magnetic fields of $10^{14} - 10^{15}$ G, whose emission is powered by magnetic field decay [38]. For such strong magnetic fields pulsar emission may be even more complicated, as these fields exceed the 'quantum critical field' strength,[5] $B_{QC} = 4.4 \times 10^{13}$ G, where higher order Quantum Electrodynamic effects play a role. For example, the amplitude for photon splitting, $\gamma \rightarrow \gamma + \gamma$, a third order effect, is proportional to $\alpha^3(\hbar\omega/m_e c^2)^5(B/B_{QC})^6$ [1] where α is the fine structure constant and $\hbar\omega$ is the photon energy. In magnetic fields $\gtrsim B_{QC}$ this dominates over photo-pair creation, quenching the build-up of plasma and hence the radio emission [4]. For many years the known magnetars were radio-quiet and the known pulsars were radio-loud, apparently well separated into groups where this process was either dominant or suppressed. However the recent discovery of three radio magnetars [7, 8, 22], J1819−1458 and a handful of other radio pulsars with $B > B_{QC}$ has changed this. While there is a dearth of pulsars in the $B \sim B_{QC}$ region the existence of any is puzzling. Weise and Melrose [37] suggest that perhaps photon splitting may not always dominates, if single polarisation selection rules forbid it. However we might ask a more basic question: if the magnetars are powered by magnetic energy, not by rotation, then why is it fair to use the vacuum rotator expression of Eq. 2.21 to estimate the magnetic field strength of magnetars? This is what is commonly done but the field strengths so derived may not be trustworthy. A related question is at what point this becomes an unreliable estimate, e.g. is it fair to use this estimate for J1819−1458? Then, in translating $P - B$ death lines to $P - \dot{P}$ space we might require a different rule than the canonical $B \propto \sqrt{P\dot{P}}$, although what that might be is not known.

[5] The quantum critical value corresponds to the energy gap of electron cyclotron orbits ('Landau levels') equalling the electron rest mass. In SI units $\Delta E = \hbar e B/m_e$ so that $B_{QC} = m^2 c^2/q\hbar$.

References

1. S.L. Adler, Ann. Phys. (USA) **67**, 599 (1979)
2. A.M. Archibald, V.M. Kaspi, M.A. Livingstone, M.A. McLaughlin, ApJ **688**, 550 (2008)
3. J. Arons, in *Pulsars: Problems and Progress*, vol. 1, ed. by S. Johnston, M.A. Walker, M. Bailes, (IAU Colloquium 160, 1996), p. 177
4. M.G. Baring, A.K. Harding, ApJ **507**, L55 (1998)
5. J. Boyles et al., in *Bulletin of the American Astronomical Society*, BAAS **41**, 464 (2010)
6. S. Burke-Spolaor, M. Bailes, MNRAS **402**, 855 (2010)
7. F. Camilo, S.M. Ransom, J.P. Halpern, J. Reynolds, D.J. Helfand, N. Zimmerman, J. Sarkissian, Nature **442**, 892 (2006)
8. F. Camilo, J. Reynolds, S. Johnston, J.P. Halpern, S.M. Ransom, ApJ **681** (2008), astro-ph/0802.0494
9. K. Chen, M. Ruderman, ApJ **402**, 264 (1993)
10. B. Davies, D.F. Figer, R. Kudritzki, C. Trombley, C. Kouveliotou, S. Wachter, ApJ **707**, 844 (2009)
11. J.S. Deneva et al., ApJ **703**, 2259 (2009)
12. B.M. Gaensler, N.M. McClure-Griffiths, M.S. Oey, M. Haverkorn, J.M. Dickey, A.J. Green, ApJ **620**, L95 (2005)
13. B.M. Gaensler, P.O. Slane, Ann. Rev. Astr. Ap. **44**, 17 (2006)
14. F.P. Gavriil, M.E. Gonzalez, E.V. Gotthelf, V.M. Kaspi, M.A. Livingstone, P.M. Woods, Science **319**, 1802 (2008)
15. P. Goldreich, W.H. Julian, ApJ **157**, 869 (1969)
16. J.W.T. Hessels, S.M. Ransom, V.M. Kaspi, M.S.E. Roberts, D.J. Champion, B.W. Stappers, in *American Institute of Physics Conference Series*, Vol. 983, ed. by C. Bassa, Z. Wang, A. Cumming, V.M. Kaspi, 40 Years of Pulsars: Millisecond Pulsars, Magnetars and More, p. 613 (2008)
17. G.H. Janssen, B.W. Stappers, R. Braun, van W. Straten, R.T. Edwards, E. Rubio- Herrera, van J. Leeuwen, P. Weltevrede, A&A **498**, 223 (2009)
18. D.L. Kaplan, P. Esposito, S. Chatterjee, A. Possenti, M.A. McLaughlin, F. Camilo, D. Chakrabarty, P.O. Slane, MNRAS **400**, 1445 (2009)
19. A. Karastergiou, A.W. Hotan, W. van Straten, M.A. McLaughlin, S.M. Ord, MNRAS **396**, 95 (2005), astro-ph/0905.1250
20. O. Kargaltsev, G.G. Pavlov in *American Institute of Physics Conference Series*, Vol. 983, ed. by C. Bassa, Z. Wang, A. Cumming & V.M. Kaspi. 40 Years of Pulsars: Millisecond Pulsars, Magnetars and More, p. 171 (2008)
21. M. Kramer, A.G. Lyne, J.T. O'Brien, C.A. Jordan, D.R. Lorimer, Science **312**, 549 (2006)
22. L. Levin et al., ArXiv e-prints (2010), astro-ph/1007.1052
23. A. Lyne, G. Hobbs, M. Kramer, I. Stairs, B. Stappers, Science **329**, 408 (2010)
24. M. McLaughlin, in *Astrophysics and Space Science Library*, Vol. 357, ed. by W. Becker, ASSL, (2009) p. 41
25. M.A. McLaughlin et al., MNRAS **400**, 1431 (2009)
26. M.A. McLaughlin et al., ApJ **670**, 1307 (2007)
27. R.P. Mignani, ArXiv e-prints (2010), astro-ph/1007.4990
28. M.P. Muno et al., ApJ **636**, L41 (2006)
29. G.J. Qiao, B. Zhang, A&A **306**, L5 (1996)
30. N. Rea et al., MNRAS (2010), in press, astro-ph/1003.2085
31. N. Rea, M. McLaughlin, in *American Institute of Physics Conference Series*, Vol. 968, Y.-F. Yuan, X.-D. Li, D. Lai, (eds.), Astrophysics of Compact Objects, 2008, p. 151
32. N. Rea et al., ApJ **703**, L41 (2009)
33. S. Reynolds et al., ApJ **639**, L71 (2006)
34. E. Rubio-Herrera, Ph.D. thesis (University of Amsterdam, 2010)
35. Y.P. Shitov, A.D. Kuzmin, D.V. Dumskii, B.Y. Losovsky, Astronomy Reports **53**, 561 (2009)

36. N. Wang, R.N. Manchester, S. Johnston, MNRAS **377**, 1383 (2007)
37. J.I. Weise, D.B. Melrose, Phys. Rev. D **73**, 045005 (2006)
38. P.M. Woods, C. Thompson, in *Compact Stellar X-ray Sources*, ed. by W.H.G. Lewin, M. van der Klis (Cambridge University Press, Cambridge, 2004), astro-ph/0406133
39. M.D. Young, R.N. Manchester, S. Johnston, Nature **400**, 848 (1999)
40. B. Zhang, A.K. Harding, A.G. Muslimov, ApJ **531**, L135 (2000)

Chapter 9
Conclusions

9.1 What Do We Know Now?

Here we quickly summarise the work presented in this thesis. After a review of radio transients (Chap. 1) and neutron stars (Chap. 2) we examined, in Chap. 3, what it would mean if RRATs were, as had been suggested, a distinct population of Galactic neutron stars. This led us to conclude that there would be a 'birthrate problem', i.e. the observed classes of neutron stars would be incompatible with the observed supernova rate.[1] However this is only the case if the classes are distinct and can be resolved if the various observed manifestations are in fact evolutionarily linked in some way. No such evolutionary framework exists for pulsars, which demonstrates our lack of knowledge of neutron star evolution post-supernova.

Although it seems clear without RRATs (e.g. see Fig. 3.2), if they could somehow be forgotten about, then the birthrate problem would be eased somewhat. So, with the RRAT estimate figuring so highly we next questioned, in Chap. 4, whether the RRAT population was in fact as large as proposed by McLaughlin et al. [25]—perhaps, for instance, there was a huge overestimation. To investigate this it is necessary to find more sources. As we had developed new algorithms and software for improved searching, we decided to completely re-process the PMPS. It had been claimed that as much as 50% of the RRATs detectable in the survey had been obscured by RFI [25] so we applied RFI mitigation techniques (see Sect. 4.4) in our re-processing. Our analysis, which we refer to as PMSingle, was successful and identified a further 19 new sources, to add to the original 11 detections. Of these 19 PMSingle sources, 12 have been observed multiple times whereas 7 have been observed on only one occasion. The sources which have not been re-observed may have very low burst rates, or, may in fact be single transient events. One source in particular is of great interest due to its suggested extragalactic distance of >50 kpc [5]. These discoveries are consistent with the initial population estimate for RRATs—we removed the effects of 'RFI blindness', which effected $\sim 1/2$ of the PMPS pointings, and (more than)

[1] See Appendix C for some supplementary information for Chap. 3.

E. F. Keane, *The Transient Radio Sky*, Springer Theses,
DOI: 10.1007/978-3-642-19627-0_9,

doubled the known PMPS RRATs. Thus the birthrate problem does not 'go away' and RRATs must be explained within the context of known neutron star classes. Fortunately, as we will discuss in the following section, this is possible.

To do this we need to further characterise their properties. With this goal, we began a focused campaign of monitoring observations so as to obtain coherent timing solutions for as many RRATs as possible through observations at Jodrell Bank and at Parkes. The methods used and difficulties encountered in this endeavour are described in Chap. 5. These studies revealed glitches in J1819−1458, with anomalous post-glitch recovery of the slow-down rate. For the original 11 RRATs the number of coherent timing solutions is now 7, up from 3. Furthermore, of the newly discovered repeating PMSingle sources, timing solutions have been obtained for 7 sources. These new solutions are presented in Chap. 6. We then described, in Chap. 7, an upcoming X-ray observation of a 'dying' pulsar, one of the PMSingle RRATs. Finally we discussed a simultaneous optical-radio observation of a RRAT, motivated by the possibility of a giant radio pulse association. Having reviewed the work of other authors in studies of transient neutron stars in Chap. 8, we now reflect on what it all means.

9.2 When a Pulsar is a RRAT

We define a RRAT as:

Definition A RRAT is a repeating radio source, with underlying periodicity, which is more easily detectable via its single pulses than in periodicity searches.

This (arbitrary) definition is clearly a detection-based definition and a source can only be labelled a RRAT for a specific survey/telescope/observing frequency/ observing time.[2] It says nothing directly about the intrinsic properties of the source—we feel that this is appropriate. Thus: *an observing setup might be contrived so as to make any pulsar a RRAT.*

Are the group of RRATs, so defined, in any way special? In a general sense, where any observational setup is possible, they are not, but for realistic survey specifications, they can be. RRAT searches[3] make a selection on the parameter space of possible sources. The group of RRATs resulting from this may be special, for a number of reasons, as we will elucidate.

9.2.1 Selection Effects

We begin by considering what this definition means as far as selection effects are concerned. As an example, we can take a source, period P, which emits (detectable)

[2] In fact the RRAT label is not permanent: a source may be detected as a RRAT but subsequently be more easily detected in periodicity observations, even for identical observing setups. This was the case for the PMSingle source J1652−4406.

[3] Here, and below, we use 'RRAT search' as a synonym for 'searches for isolated single bursts'.

pulses a fraction of the time g and nulls (or is not detected) a fraction of the time $1-g$. Then we can use the resultant selection effect in $g-P$ space for this scenario, derived earlier in Sect. 4.2, Eq. 4.22, namely $r > 1$ when $Tg^2/4 < P < Tg$, where T is the observing time. For a given g, the low period limit defines the $r = 1$ condition, so that, at lower periods an FFT search is more effective. For higher periods than Tg there is unlikely to be even one pulse during the observation. Figure 9.1 shows a plot of $g-P$ space with 'RRAT-PSR' boundaries marked for the 35-min pointings of the PMPS. Here we are using our definition of 'RRAT', and using 'pulsar/PSR' as a synonym for 'more easily, or only detectable in a periodicity search'. Thus PMPS RRATs are those sources in the orange or white shaded regions. Different surveys will have different RRAT-PSR boundaries, e.g. the higher-latitude Parkes surveys analysed by Burke-Spolaor and Bailes [2] had shorter pointings and hence different boundaries which are over-plotted on Fig. 9.1. Thus the 'RRAT' J1647−36 detected in the high-latitude surveys would have been detected as a 'pulsar' if it were surveyed in the PMPS. We note that, in reality, the g values we measure represent the *apparent* nulling fraction, i.e. the intrinsic values of g may be higher depending on the pulse-to-pulse modulation and distance to the source [2, 34]. Periodicity searches also make a selection in $g-P$ space, the grey region of Fig. 9.1. In comparison to periodicity searches, RRAT searches are sensitive to high period sources ($\gtrsim$10 s) with moderate nulling fraction ($\sim$0.1) down to very short period ($\sim 10^{-3} - 10^{-1}$ s) sources with large nulling fraction ($10^{-4} - 10^{-3}$).

From inspecting Fig. 9.1, we can make a number of remarks. Firstly, we can see that the average 'RRAT' and 'PSR' periods we infer would be:

$$\langle P\rangle_{\rm RRAT} = \frac{\int P(\int RRAT(g,P)dg)dP}{\int\int RRAT(g,P)dgdP}, \quad \langle P\rangle_{\rm PSR} = \frac{\int P(\int PSR(g,P)dg)dP}{\int\int PSR(g,P)dgdP}, \tag{9.1}$$

where $RRAT(g,P)$ and $PSR(g,P)$ are distribution functions in $g-P$ space. For a uniform $g-P$ distribution these simply correspond to the areas of the orange and grey regions in the figure, and the results can be easily calculated. For sensible ranges (the ranges plotted in Fig. 9.1, for $P < 10$ s, say) we always get $\langle P\rangle_{\rm RRAT} > \langle P\rangle_{\rm PSR}$. It would not then be useful (or fair) to compare period distributions of sources selected in these ways. Further examining the figure we can see that the bottom left-hand corner (bounded by the blue lines in the figure) is lacking in sources. Moving upwards a decade in P for the same g range (say) we expect to get $\sim$10 times as many sources, if the distribution is uniform, and this is, roughly, what we see. Going up another decade in P we do not see a further increase in sources, most likely due to there being no radio-visible pulsars with $P \gtrsim 10$ s. The period distribution seems approximately uniform in $\log P$ in the band $\sim$0.5–8 s, although, given the small numbers of sources, it is not inconsistent with a lognormal distribution centred at $\sim$3 s, which we contrast with the lognormal distribution for pulsars centred at 0.3 s [29].

The distribution in g may be of more interest. We can see that, within the band where we see sources, that the distribution is not uniform. If anything the distribution looks uniform in $\log g$. We can thus explain the distribution of sources as follows:

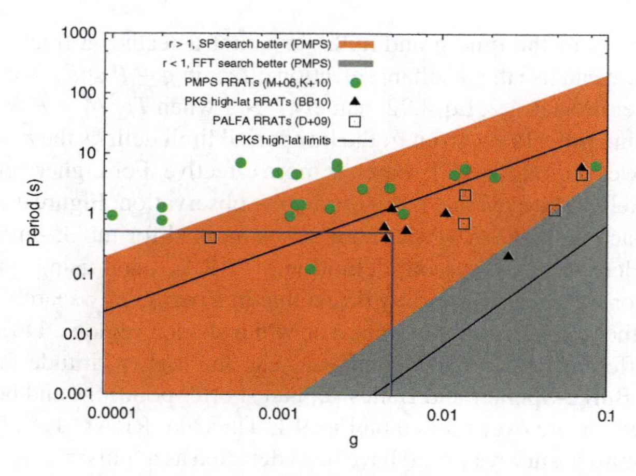

Fig. 9.1 Plotted is $g - P$ space with the regions where SP searches (*orange*) and FFT searches (*grey*) are more effective for the PMPS [22], defining "RRAT" and "pulsar" regions. Over-plotted are the PMPS RRATs with measured periods as reported in Keane et al. [14], McLaughlin et al. [25] (M+06 and K+10 in the figure), and described in Chaps. 4, 5 and 6. Also plotted are the boundaries (*black lines*) for the sources reported by Burke-Spolaor and Bailes [2] (BB10 in figure) with known P and g. We also plot the sources reported in Deneva et al. [6] (D+09 in the figure). J1854+0306 is plotted with the PMPS sources, although it was also identified in PALFA. We note that the boundaries for the inner-Galaxy PALFA pointings are the same as for the Parkes high-latitude surveys if we assume no difference in sensitivity. This is of course incorrect, and due to this extra difference (the Parkes surveys have the same sensitivity as each other) the D+09 sources are plotted simply for illustration

(i) the low P–low g region is devoid of sources as this does not represent a large area of parameter space and/or there does not seem to be many sources with these characteristics; (ii) the $P \gtrsim 10$ s region does not have any active radio pulsars, consistent with what is expected for slow pulsars which have passed the death line; (iii) the $P \sim 0.5$–8 s region for $g \sim 10^{-4} - 10^{-1}$ shows a somewhat uniform distribution in $\log g$, suggesting that there are more RRAT-selected pulsars with high nulling fractions than would be expected from a uniform distribution in $g - P$ space. To turn this around, if we search for RRATs, it seems that we are likely to find pulsars with high nulling fraction. These data are not sufficient to identify any trend in g with P, and there are less data for investigating any relationships with $\dot{P}$, τ, B, $\dot{E}$, etc. As a final comment on Fig. 9.1 we note that there are several PMPS sources just above the orange region. Here it is unlikely that there will be a pulse during a 35-min pointing but nevertheless there are 8 sources. For each of these, which we were lucky to detect, we might expect there are several similar sources, which we missed, simply due to bad luck.[4]

As we have described in detail in Chaps. 5 and 6, monitoring RRATs over some time reveals their slow-down rate, $\dot{P}$. This is not subject to any selection effect in either RRAT or periodicity searches, as typical $\dot{P}$ values have no effect during

[4] This is yet another argument, if any were needed(!) in favour of surveying the sky multiple times.

survey pointings. Looking at Fig. 6.5 we can see that the $\dot{P}$ values for 5 RRATs in particular seem higher than average, with high corresponding magnetic field strengths of $B \gtrsim 10^{13}$ G (using Eq. 2.21). Bar J1554−5209 the other sources all have slightly higher than average magnetic fields with $B > 10^{12}$ G, consistent with the earlier claim of McLaughlin et al. [24]. As RRAT searches select high nulling fraction pulsars, and these same sources seem to have high-B values, this suggests the question: Do long period and/or high-B pulsars have higher nulling fraction? Here we reach a dead end as the nulling properties of pulsars are completely unknown in the $B \sim B_{\rm QC}$ and $P \gtrsim 3$ s regions, where a number of RRATs are found. One reason for this is that these regions have a dearth of sources and in fact the PMPS RRATs represent a significant fraction of the known sources in these regions. As the PMPS RRATs are not obviously very distant we also ask the question: Do long period and/or high B pulsars have large modulation indices? Weltevrede et al. [34, 35] suggest a weak correlation of modulation index with B, but again, the number of high-B and long period sources in this sample was small.

We mentioned another selection effect that the PMPS RRATs suffered from—the 'low-DM blindness' of the original RRAT search, i.e. the possibility that low-DM sources were missed due to the effects of RFI. Of course, it is difficult to determine how many sources would be missed, because, as the DM increases the Galactic volume searched increases, but in a non-steady way as DM is only a rough proxy for distance (see Fig. 4.13). Our re-analysis removed this effect and in fact discovered a number of low-, as well as high-DM sources which had initially been missed due to RFI (see e.g. Fig. 4.8; Table 4.4) so that we believe this selection effect has been removed.

9.2.2 'Solutions'

As we described in Sect. 2.3.2 there have been many 'solutions' proposed to what RRATs might be. However, as we have asserted that RRATs are merely pulsars which fit a particular selection criteria, for a given observational setup, the question of a 'solution' becomes more a question of what types of pulsars are we most likely to detect as RRATs. There are two obvious types ('solutions' if you will), consistent with high observed nulling fractions: (i) weak/distant pulsars with high modulation indices; (ii) nulling pulsars.

The high projected population of RRATs can be absorbed somewhat if some sources are covered by solution (i). Such sources will have low-luminosity periodic emission. The pulsar population is estimated only above some threshold luminosity, typically $L_{\rm min} \sim 0.1$ mJy kpc^2 (see Sect. 3.3), so that if these sources are above $L_{\rm min}$ they are already accounted for within low-luminosity selection-effect scaling factors in estimates of the pulsar population (see e.g. [20, 29]). If the underlying periodic emission were below $L_{\rm min}$ then these sources would contribute to a birthrate problem by increasing the pulsar population estimate, and indeed the required low-luminosity turn-over[5] is not yet seen, which is why artificial cut-offs are usually applied in population syntheses (see e.g. Faucher-Giguère and Kaspi [9]). Burke-Spolaor and Bailes [2] argue that extreme modulation can account for all but two RRATs, but notably not J1819−1458 (and J1317−5759), which agrees with Miller et al. The true number covered by scenario (i) may be smaller as it assumes analogues of the extreme source PSR B0656+14 to be common in the Galaxy. So it seems that a number of RRATs[6] are accounted for by scenario (i), whereas some are not, and seem to fit type (ii).

We note that Burke-Spolaor and Bailes [2] have dubbed "objects which emit only non-sequential single bursts with no otherwise detectable emission at the rotation period", as 'classic RRATs'. By this definition, there may be no RRATs (see discussion of misconceptions in Sect. 9.3.1). In our view, such extra labelling is unhelpful[7] and we do not use it.

9.2.3 Switching Magnetospheres?

Scenario (ii), which sees RRATs as nulling pulsars, extends the boundaries of observed nulling behaviour. In comparison to the previously observed nulling sample, RRATs would be considered extreme, with nulls of minutes to hours, as opposed to ~seconds. Excluding the RRATs, nulling has been observed in ~50 pulsars, but, if we include pulsars where an upper limit on the nulling fraction has been obtained, the number in the literature is ~100 [1, 10, 19, 28, 32–35]. Of these, there are 50 with $P>1$ s, 10 with $P>2$ s and 1 with $P>3$ s. The nulling behaviour of long period and high-B sources is completely unknown. Some authors have claimed a correlation of nulling fraction with period [1], whereas others have claimed the correlation is instead with characteristic age [33]. Some of the observed RRATs are high-B sources with long period, but are young in terms of characteristic age. Others are 'dying' pulsars having both long periods and old characteristic ages. Observations of a large sample of pulsars, selected as RRATs, could then be ideal for the purpose of testing these competing claims.

Thus, we have 'nulling pulsars', with nulls of 1–10 periods, 'RRATs' with nulls of $10 - 10^4$ periods and 'intermittent pulsars' with nulls of $10^4 - 10^7$ periods. It seems like there may be a continuum of null durations in the pulsar population. The question of the 'RRAT emission mechanism' is then subsumed by the questions of what makes pulsars null, and why such a wide range of null durations are possible. Another question of immediate interest is in what cases do nulls occur—high-B, long period, old pulsars? Also unexplained are the non-random [27] and periodic

[5] There must be a low-luminosity turn-over so that the integral $\int N(L)dL$ does not diverge at the low end. Here $N(L)dL$ denotes the number of pulsars with luminosity between L and $L+dL$.

[6] RRAT pulse amplitude distributions will shed more light on these matters (Miller et al. submitted).

[7] It might add to the belief that RRATs are a completely distinct population.

behaviour seen in several sources, e.g. 1-min periodicity for PSR J1819+1305 [26], several minutes for the PMSingle source J1513−5946 (see Sect. 4.5.2), hours for PSR B0826−34 [7], ~1 day for PSR B0823+16 (N. Young, private communication), ~1 month for PSR 1931+24 [18] and perhaps several months for PSR J1809−15 [8]. Considering the more general case of moding, we can add the pulsars reported by Lyne et al. [21], which switch between (at least) 2 modes, with associated switches in spin-down rate. Another recently discovered source—PSR J0941−39 is observed to switch between 'RRAT-like' and 'pulsar-like' modes [2], and there is at least one other pulsar known to switch between RRAT, pulsar and null states (P. Weltevrede, private communication). All of this mounting evidence suggests that it is a general property of (at least some) pulsars, that they can switch back and forth between two stable states of emission, and that nulling fraction may evolve in steps, rather than continuously.

We make the important note that, as $\dot{E}_{\rm radio} \ll \dot{E}$, the simple switching on or off of the radio emission should not result in any noticeable effect[8] in $\dot{\nu}$. The fact that $\dot{\nu}$ changes have been observed in very-log duration nullers (Kramer et al. [18], the effect is not observable in short-duration nullers [15]) suggests a large-scale change in the magnetosphere, i.e. the nulls are not due to the micro-physics of the emission mechanism [31]. Within the framework of force-free magnetospheres, it has been shown that a number of stable solutions are possible with different sizes of the closed field line region [3, 30]. These solutions are derived as for the Contopoulos et al. [4] solution discussed in Sect. 2.3.1, but without the assumption that the angular velocity of the field lines is equal to that of the star. Timokhin [31] has shown how moderate changes in the beam shape and/or current density can cause large changes in $\dot{E}$, and hence $\dot{\nu}$. For a pulsar changing between two stable states, the observed emission along our line of sight will change, and this will be seen as a mode switch. A null will result if the beam moves out of our line of sight as a result of the switch, or, if there is a sufficient change in current such that the emission ceases [31].

What seems clear from the data is that pulsars can switch between stable states. Such an effect, if truly a generic property of pulsars, can explain the phenomena of moding, nulling and RRATs. The theoretical work shows that different stable magnetospheric states exist. The reason why a pulsar would switch between two states (in particular with a periodicity) is unknown. Contopoulos [3] have shown that a sudden depletion of charges will result in such a change of state (which they refer to as a 'coughing magnetosphere'), but with no explanation for why this depletion might arise.

---

[8] Consider a simple calculation for a pulsar with radio flux density of 10 mJy, a distance 1 kpc away. Its radio pseudo-luminosity is then 10^{-2} Jy kpc^2 $\approx 10^11$ W Hz^{-1} (see Eq. A.9). Assuming a constant flux density over a GHz bandwidth gives a luminosity of $E_{\rm radio} = 10^{20}$ J s^{-1} $= 10^{27}$ erg s^{-1} which we can compare to the $\dot{E}$ values reported throughout this thesis.

9.2.4 The PMPS RRATs

The observed PMPS RRATs seem to fit loosely into three classes. These are: (1) high-B pulsars with $B \sim B_{QC}$; (2) 'normal' pulsars; and (3) 'dying' pulsars hovering near the pulsar death line.

The high-B sources are of interest as they occupy an empty region of $P - \dot{P}$ space, between the radio pulsars and the magnetars. It is in this region that radio emission should stutter and fail so the ability to identify such sources in RRAT searches is an important result of the recent transient searches. J1819−1458 occupies this region, and as we have discussed in Sect. 5.2.2, it undergoes anomalous glitches. We plan to monitor all of the high-B sources, and indeed search for more, to investigate what the significance of these strange glitches is, and how, if at all, they relate to transient/switching behaviour which is observed.

The dying pulsars provide information on radio emission in old neutron stars, and, the identification of more is important in challenging proposed death lines. This is just one aspect of the work required to identify the elusive radio pulsar emission mechanism. Furthermore, if we are lucky, and identify a nearby source, such as J1840−1419, we can investigate thermal characteristics of neutron stars. The continued study of all these classes of RRATs, incorporated into a wider study of high-B and long period pulsars, will provide valuable observational data for understanding the apparently common phenomena of moding and nulling.

We re-iterate that there is no reason to consider any of these classes as a distinct population, nor is there a need to formulate any new emission mechanisms. They can be explained within the existing pulsar framework, or rather, the existing framework of open questions. What is interesting is that, with RRAT searches, we have a means with which to identify pulsars which have been difficult to find, in particular the high-B and the dying pulsars.

9.3 Questions Answered

Here we address the questions posed in Sect. 2.3.2:

(1) Are there really as many RRATs in the Galaxy as the initial estimates imply? How well do we know the parameters in Eq. 2.37? There does seem to be many pulsars which would be detected as RRATs in surveys such as the PMPS. We now have essentially removed the f_{RFI} factor from Eq. 2.37 but, as we have discussed, the beaming and burst rate distributions are also vital ingredients in a population estimate. We note that the population estimate is not to be thought of as representing a distinct group. Interestingly, nulling is not considered when performing population syntheses (basically due to lack of information), even though the discovery of so many nulling sources of late indicates that this may be an important detail to include in evolutionary models.

(2) Are they truly a distinct population? What are the implications of this? RRATs are not a distinct population of neutron stars. We have shown that if this were true then the implied Galactic population of neutron stars would be too large to be accounted for by the supernova rate. The discovery of 19 extra sources in the PMPS also retracts from any idea that the RRATs are a distinct population, but less abundant than previously thought. We propose that RRATs are in fact simply radio pulsars. Although we think of 'RRAT' as a detection label, the possibility remains that those pulsars (or a subset thereof), discovered as RRATs, may represent an evolutionary state with a high associated nulling fraction.

(3) Why do they have longer periods than the radio pulsars? Is this significant? When we search for RRATs we make a selection in $g - P$ space which favours the detection of (apparently) nulling pulsars. These searches select high period sources, but we do remain sensitive to short periods (see Fig. 9.1). The significance of detecting long period sources may be that it indicates an increased nulling fraction and/or increased pulse-to-pulse modulation for long period/old neutron stars.

(4) What decides whether a NS will manifest itself as a RRAT, as opposed to (say) a magnetar or an XDINS, which occupy the same region of $P - \dot{P}$ space? We do not yet know the answer to this important question, although the answer is fundamental if there is to be "grand unification of neutron stars" [13], i.e. the determination of some kind of evolutionary framework. The region of $P - \dot{P}$ space defined by $P = 4{-}10$ s, $\dot{P} = 10^{-13}{-}10^{-12}$ contains radio pulsars (some 'normal' pulsars, some RRATs like J1819−1458), magnetars and XDINSs. For very similar spin-down properties we have very different observational manifestations. We might speculate[9] that these different classes, although having similar properties now, have evolved in completely different ways and may have completely different ages. The conditions for coherent radio emission may be very sensitive, with this region a particular area of parameter space on the threshold for emission. This is perhaps consistent with the transient radio emission seen in magnetars and the extreme nulling of the RRATs in this region. If the re-connection rate at the Y-point were slow, or progressed in steps, then bursts of radio emission may be expected between dormant phases, when the magnetospheric configuration was favourable. Regarding the XDINSs, it has been suggested that they may exhibit radio emission but suffer from unfavourable beaming (see e.g. Kondratiev et al. [16, 17]), but as their spectra seem to be purely thermal this may not be the case [11]. Only the discovery of more XDINSs can settle this question convincingly.

(5) Are their observed properties a result of selection effects in our search methods or truly a representation of a class of neutron stars? Given the parameters of our survey and searches, are these the kind of sources we expect to find? We have already discussed the selection in $g - P$ space, but in addition, although there is no selection effect, it seems that sources with high $\dot{P}$ (and thus high-B)

[9] The remainder of the answer to this question is conjecture!

are selected. The significance of this is that high-B and/or long-period pulsars are suggested to have either a high null fraction or stronger pulse-to-pulse modulation.

(6) How different is their emission in comparison to the radio pulsar population? From the currently available data, their radio emission seems to be the same as that seen from pulsars. The only difference seems to be the sporadicity of the emission, implying large nulling fraction and/or pulse-to-pulse modulation. The polarisation properties of J1819$-$1458 are not unusual for its value of $\dot{E}$ [12]. There have not yet been any baseband studies, at the highest time resolution, however the evidence available (see e.g. Chap. 7) suggests that their emission is not like that from GRP sources. The X-ray properties are just beginning to be investigated—J1819$-$1458 has been observed multiple times, but only non-detections have been obtained, so far, for the other four sources studied in this band. Very recently, there have also been optical, infrared and γ-ray attempts made, as we have reported in Chaps. 5, 7 and 8.

(7) What are there long-term timing properties? How stable, or not, are these? As discussed in Chap. 6, there are now long-term timing solutions for 14 PMPS RRATs and Fig. 6.4 shows their distribution in $P - \dot{P}$ space. For some of the original sources, ~6 years of timing observations are available. For the PMSingle sources this is ~1.5 years. J1819$-$1458 has shown anomalous glitches, whose significance we are yet to understand, but the other sources have so far shown stable timing properties. In fact some sources, in particular the long-period sources, show remarkably stable timing solutions, especially considering the method of timing via single pulses (see Chap. 5). Monitoring of all sources is ongoing at several telescopes.

(8) Are they old or young? Are they evolutionarily linked to any of the previously known classes of neutron star? There is no single answer for this question regarding age, which covers all of the RRATs. This is perhaps because the question is suggestive that all of the RRATs are a distinct group or evolutionary stage when in fact we see a number of 'solutions'. As we have said, the characteristic ages of the PMPS RRATs are not remarkable in comparison to the overall pulsar population (see Fig. 6.5), but we can comment on the three groups that these sources seem to lie in. Bar their sporadicity, the RRATs amongst the normal pulsars seem to have no remarkable properties. The high-B sources are apparently young, by the possibly very unreliable measure of characteristic age. The RRATs near the death line are apparently old, by the same measure, something which upcoming X-ray studies of J1840$-$1419 may shed some further light on.

(9) Can we characterise the observed properties more completely? Are more timing solutions possible and where in $P - \dot{P}$ space do RRATs really live? Through our monitoring observations over the past 3 years we have been able to characterise the observed radio properties of the known RRATs. In particular, the known timing solutions have increased to 14, up from three. Chapters 5 and 6 discuss what has been observed, in detail.

(10) Can we discover new sources and improve the characterisations to help to answer all the above questions and identify any key relationships? We have discovered 19 new sources, 12 of which have been observed on multiple occasions as part of a followup campaign of monitoring over the last ~1.5 years. As we have discussed in this chapter, the RRAT searches seem to select highly modulated and/or extreme nulling pulsars. The observed PMPS RRATs can be roughly grouped as: four (or perhaps six) seem to be 'normal' pulsars, four seem to be high-B radio pulsars (one of which is above the photon-splitting line) and four are old pulsars, some of which are quite close to the death line.

9.3.1 Facts Abouts RRATs

We now address a number of assertions, claims and misconceptions concerning the characteristics of RRATs, that we have encountered during the last 3 years, which are held to be correct. Some of these are true, some of these are not.

All RRATs are high-B and therefore linked to magnetars in some way. FALSE. If we arbitrarily define high-B as $B \geq 10^{13}$ G, then there are 5 RRATs which have high magnetic fields. Of those, J1819−1458, with $B = 5 \times 10^{13}$ G, remains the RRAT with the strongest magnetic field strength. Besides the tentative link suggested for J1819−1458, due to its unusual glitches, it is certainly not true to say that links have been identified between the other RRATs and magnetars.

RRATs are only detectable in single pulse searches. FALSE.[10] Several of the original and PMSingle sources are detectable in periodicity searches, in some cases occasionally and in some cases reliably.

The arrival times of RRAT pulses are random. FALSE (see footnote 10). We have discussed non-random behaviour in the PMSingle sources, where clustering of pulses is seen, e.g. J1724−35 and J1513−5946 (Sect. 4.5.2). This is also seen in J1913+1330 at Jodrell (Sect. 5.2.3) and at Parkes [23]. Recent work by Palliyaguru et al. (in preparation) has also shown this in six of the eight original RRATs.

Consecutive pulses from RRATs are never seen. FALSE (see footnote 10). One implication of the non-random distribution of pulse amplitudes seen in the Jodrell (see Chap. 5) and Parkes (Paliyaguru et al., in preparation, see Sect. 8.1) observations are that we might not see isolated pulses from RRATs. In fact, Palliyaguru et al. report just this—observing higher instances of doublets, triplets and quadruplets, particularly for J1819−1458, than would be expected by random chance, in the Parkes data for the original RRATs. We can confirm that this is seen in the Jodrell Bank observations of J1819−1458 and J1913+1330 as well as the Parkes observations of the PMSingle sources, as described throughout Chaps. 4, 5 and 6. Intriguingly, Palliyaguru et al. also report an instance of detecting pulses from J1819−1458 for nine consecutive periods. This drastically changes the 'activation timescales' needed in some

[10] This assertion was true of the data which had been accumulated up to the original discovery paper [25], but is no longer a true statement.

models (although not all, see e.g. Zhang et al., 2007) of RRAT emission, from ~3 ms to ~35 s.

The RRATs are all isolated neutron stars. TRUE. The RRATs discovered so far are all isolated (although see the discussion on the possible origins of J1846−0257 in Sect. 8.1) but this is not in any way a defining feature.

RRATs are special. TRUE. Although this may depend on who you ask.

9.4 Future Work

Throughout the research presented in this thesis, questions have been answered, but, as a result, several more have been raised. The result is a number of lines of enquiry, which we suggest, if followed up by researchers in the field, may provide some useful information in attempts to understand transient radio neutron stars and pulsars in general.

- What is the significance of the anomalous glitches in J1819−1458? Are these common in this source? Are they seen in any other sources?
- What are the γ-ray properties of the extreme pulsars identified as RRATs? A thorough examination of all sources would be useful.
- What are the X-ray properties of the RRATs? Can this tell us anything about their evolution?
- What are their long-term (i.e. over ≥5 years) timing properties?
- With the discoveries of neutron stars which switch between 2 or more stable states, it seems timely, and necessary, to perform a complete census of nulling pulsars across the $P - \dot{P}$ diagram. Nulling properties are known for only a relatively small fraction of the pulsar population, and not at all for the high-B and long-period sources.
- What causes nulling?
- What dictates the onset of nulling in a pulsar? When it slows down below a critical rotation rate? When its surface magnetic field grows/decays to a certain value?
- What are the properties of pulsars occupying 'void' regions of $P - \dot{P}$ space?
- Can force-free magnetospheric models produce a set of stable solutions with a natural explanation for the periodic switching between such states, which seems to be observed?
- How many neutron stars are there in the Galaxy? The search for occasional bursts has enabled us to identify a source at the far reaches of our Galaxy (J1852−08). How many such sources are there?
- How many RRATs and radio transient bursts are there in the numerous pulsar surveys yet to be searched for isolated bursts?
- If spin-down properties are not enough to uniquely define how a source will appear to us (i.e. what wavelength, steady or bursting) then what are the vital diagnostics which do?
- What are the polarisation properties of the RRATs?

- What are the radio spectral properties of the RRATs? And how do these evolve with time?
- What else don't we know?!

References

1. J.D. Biggs, ApJ **394**, 574 (1992)
2. S. Burke-Spolaor, M. Bailes, MNRAS **402**, 855 (2010)
3. I. Contopoulos, A&A **442**, 579 (2005)
4. I. Contopoulos, D. Kazanas, C. Fendt, ApJ **511**, 351 (1999)
5. J.M. Cordes, T.J.W. Lazio, preprint (2002). arXiv:astro-ph/0207156
6. J. S. Deneva et al., ApJ **703**, 2259 (2009)
7. J.M. Durdin, M.I. Large, A.G. Little, R.N. Manchester, A.G. Lyne, J.H. Taylor, MNRAS **186**, 39 (1979)
8. R.P. Eatough, Ph.D. thesis (University of Manchester, 2009)
9. C.-A. Faucher-Giguère, V.M. Kaspi, ApJ **643**, 332 (2006)
10. A.J. Faulkner et al., MNRAS **355**, 147 (2004)
11. F. Haberl, Astrophys. Space Sci. **308**, 171 (2007) astro-ph/0609066
12. A. Karastergiou, A.W. Hotan, van W. Straten, M.A. McLaughlin, S.M. Ord, MNRAS **396**, 95 (2009) (astro-ph/0905.1250)
13. V.M. Kaspi, ArXiv e-prints (2010). astro-ph/1005.0876
14. E.F. Keane, ArXiv e-prints (2010). astro-ph/1008.3693
15. E.F. Keane, D.A. Ludovici, R.P. Eatough, M. Kramer, A.G. Lyne, M.A. McLaughlin, B.W. Stappers, MNRAS **401**, 1057 (2010)
16. V.I. Kondratiev, M. Burgay, A. Possenti, M.A. McLaughlin, D.R. Lorimer, R. Turolla, S. Popov, S. Zane, in *American Institute of Physics Conference Series*, ed. by C. Bassa, Z. Wang, A. Cumming, V.M. Kaspi, 40 Years of Pulsars: millisecond Pulsars, Magnetars and More, vol. 983, (2008), p. 348
17. V.I. Kondratiev, M.A. McLaughlin, D.R. Lorimer, M. Burgay, A. Possenti, R. Turolla, S.B. Popov, S. Zane, ApJ **702**, 692 (2009)
18. M. Kramer, A.G. Lyne, J.T. O'Brien, C.A. Jordan, D.R. Lorimer, Science **312**, 549 (2006)
19. D.R. Lorimer, F. Camilo, K.M. Xilouris, ApJ **123**, 1750 (2002)
20. D.R. Lorimer et al., MNRAS **372**, 777 (2006)
21. A. Lyne, G. Hobbs, M. Kramer, I. Stairs, B. Stappers, Science **329**, 408 (2010)
22. R.N. Manchester et al., MNRAS **328**, 17 (2001)
23. M. McLaughlin, in *Astrophysics and Space Science Library*, ed. by W. Becker, vol. 357 (ASSL, 2009), p. 41
24. M.A. McLaughlin et al., MNRAS **400**, 1431 (2009)
25. M.A. McLaughlin et al., Nature **439**, 817 (2006)
26. J.M. Rankin, G.A.E. Wright, MNRAS **385**, 1923 (2008)
27. S.L. Redman, J.M. Rankin, MNRAS **395**, 1529 (2009)
28. S.L. Redman, G.A.E. Wright, J.M. Rankin, MNRAS **357**, 859 (2005)
29. J.P. Ridley, D.R. Lorimer, MNRAS **404**, 1081 (2010)
30. A.N. Timokhin, MNRAS **368**, 1055 (2006)
31. A.N. Timokhin, ArXiv e-prints (2009). astro-ph/0912.2995
32. M. Vivekanand, MNRAS **274**, 785 (1995)
33. N. Wang, R.N. Manchester, S. Johnston, MNRAS **377**, 1383 (2007)
34. P. Weltevrede, R.T. Edwards, B.W. Stappers, A&A **445**, 243 (2006)
35. P. Weltevrede, B.W. Stappers, J.M. Rankin, G.A.E. Wright, ApJ **645**, L149 (2006)

Appendix A
Basic Equations of Radio Astronomy

A telescope observing a cosmic source measures an energy dE per unit area dA per unit band-width $d\nu$ per unit solid-angle $d\Omega$ st

$$dE = I_\nu dA dt d\nu d\Omega, \tag{A.1}$$

where I_ν is the 'specific intensity' (in units of $\mathrm{W\,m^{-2}\,Hz^{-1}sr^{-1}}$) is the property which is fundamental to the source. Everything else depends on us—telescope size, band-width, etc. Radio astronomers preferentially refer to the 'flux density' S_ν (in units of $\mathrm{W\,m^{-2}\,Hz^{-1}}$) which is given by:

$$S_\nu = \int_{\Omega_{\mathrm{source}}} I_\nu d\Omega, \tag{A.2}$$

where the density is with respect to frequency such that if you integrate over a frequency band B you get the flux S incident on the telescope in that bandwidth

$$S = \int_B S_\nu d\nu. \tag{A.3}$$

Confusingly the subscript is usually dropped when referring to flux density so that it is usually donated simply as S. To make matters worse, it is also commonly referred to (incorrectly) as simply the 'flux'. If this is the case however the true meaning is usually obvious from the context and/or the units. Note the common radio astronomy unit of the Jansky as a measure of flux density which is defined as:

$$1\,\mathrm{Jy} = 10^{-26}\,\mathrm{W\,m^{-2}\,Hz^{-1}}. \tag{A.4}$$

The specific intensity for a blackbody is given by Planck's Law which is:

$$I_\nu = \frac{2h\nu^3}{c^2}\frac{1}{e^{\frac{h\nu}{kT}} - 1}. \tag{A.5}$$

E. F. Keane, *The Transient Radio Sky*, Springer Theses,
DOI: 10.1007/978-3-642-19627-0, © Springer-Verlag Berlin Heidelberg 2011

For radio astronomy wavelengths we know that $h\nu \ll kT$, i.e. we are on the Rayleigh–Jeans tail of the blackbody curve, so that $e^{\frac{h\nu}{kT}} = \sum \left(\frac{h\nu}{kT}\right)^n / n! \approx 1 + \frac{h\nu}{kT}$ and then I_ν approximates to:

$$I_\nu = \frac{2\nu^2 k T_\mathrm{B}}{c^2}, \tag{A.6}$$

where T_B is the brightness temperature of the source, the black-body temperature required to have a specific intensity of I_ν. The brightness temperature is commonly used as a measure of luminosity of radio sources even when the emission is non-thermal and in fact for the majority of the transient radio emission discussed in this thesis the origin is non-thermal. As radio astronomers preferentially use flux density, S, as their standard unit of 'brightness' it is more usual to express T_B like:

$$T_\mathrm{B} = \frac{S\lambda^2}{2k\Omega}, \tag{A.7}$$

where Ω is the solid angle covered by the source and k is Boltzmann's constant. The solid angle is given by $\Omega = \pi R^2 / D^2$ and we know that the size of the emitting area must be less than the light travel st $2R < cW$ for an observed pulse width W. Substituting these expressions we get:

$$T_\mathrm{B} = \left(\frac{4}{\pi}\right) \frac{SD^2}{2k(\nu W)^2} = \left(\frac{4}{\pi}\right) \frac{L}{2k(\nu W)^2}. \tag{A.8}$$

Here $L = SD^2$ is referred to as the 'radio luminosity' but once again we note that it might more properly have a frequency subscript as its units are $\mathrm{W\,Hz^{-1}}$. There are some useful unit conversion factors we can make a note of, such as:

$$1\,\mathrm{W\,Hz^{-1}} = 1.05026 \times 10^{-13}\,\mathrm{Jy\,kpc^2}, \tag{A.9}$$

$$1\,\mathrm{erg\,s^{-1}\,Hz^{-1}} = 1.05026 \times 10^{-20}\,\mathrm{Jy\,kpc^2}, \tag{A.10}$$

and in practise these are the more usually used units. This gives us an expression for brightness temperature of the form:

$$T_\mathrm{B} \geq 4 \times 10^{17}\,\mathrm{K} \left(\frac{\mathrm{SD^2}}{\mathrm{Jy\,kpc^2}}\right) \left(\frac{\mathrm{GHz\,s}}{\nu \mathrm{W}}\right)^2. \tag{A.11}$$

Note that sometimes this is quoted with an extra factor of 2. This is because the denominator factor of 2 carried by many of the above equations (from the Planck Law) is removed when dealing with just one sense of polarisation (i.e. when there is 100% polarised emission). For unpolarised emission, both senses of polarisation contribute equally to the flux density and hence the factor of 2 (T. Bastian, private communication). Also, rather confusingly, the $4/\pi$ factor is sometimes approximated to 1 (as in Osten and Bastian [2]) so that the leading factor in this expression for the brightness temperature can be either 4, 6 or 8.

The modified radiometer equation usually used for folded observations of pulsars is [1]:

$$S = \frac{GT_{\rm sys}}{\sqrt{n_{\rm p}\Delta\nu T}}\sqrt{\frac{\delta}{1-\delta}}(S/N), \tag{A.12}$$

where $T_{\rm sys}$, $n_{\rm p}$ is the number of polarisations summed, $\Delta\nu$ is the bandwidth, T is the observing time and δ is the pulse duty cycle. G is the telescope gain in units of Jy $\rm K^{-1}$, although sometimes the reciprocal of this is referred to as the gain. Sometimes the sensitivity of a telescope is quoted in units of $\rm m^2\,K^{-1}$, which refers to the quantity $A_{\rm eff}/T_{\rm sys}$, where $A_{\rm eff}$ is the effective area of the telescope. This is related to the gain by $G = 2k/A_{\rm eff}$ so that the radiometer equation can also be expressed as:

$$S = \frac{2kT_{\rm sys}}{A_{\rm eff}\sqrt{n_{\rm p}\Delta\nu T}}\sqrt{\frac{\delta}{1-\delta}}(S/N). \tag{A.13}$$

Appendix B
Neutron Stars: Supplementary

B.1 Minimum Neutron Star Radius

The TOV equations can be re-arranged to give (see e.g. [6]):

$$\frac{P}{\rho}=\frac{(1-2Mr^2/R^3)^{1/2}-(1-2M/R)^{1/2}}{3(1-2M/R)^{1/2}-(1-2Mr^2/R^3)^{1/2}}, \tag{B.1}$$

where we have set $G=c=1$ for convenience. We can determine a minimum radius if we require that the central pressure be finite. For the case of constant ρ, and at centre of the star (i.e. $r=0$), this expression becomes:

$$\frac{P}{\rho}=\frac{1-\sqrt{1-x}}{3\sqrt{1-x}-1}, \tag{B.2}$$

where $x=2M/R$. If $P(r=0)<\infty$ then $3\sqrt{1-x}-1>0$ and thus:

$$R_{\mathrm{min}}=\frac{9}{8}R_{\mathrm{S}}, \tag{B.3}$$

where $R_{\mathrm{S}}=2M$ is the Schwarzschild radius.

B.2 Neutron Star Internal Structure

In Sect. 2.2.2, we gave the parametric EoS for a degenerate Fermi gas of electrons or neutrons, Eqs. 2.9 and 2.10. The constants in these equations are given by (see e.g. [7]):

$$A_{\mathrm{e}}=\frac{m_{\mathrm{e}}^4c^5}{\hbar^3}=\frac{m_{\mathrm{e}}c^2}{\lambda_{\mathrm{e}}^3}=1.42\times10^{24}\ \mathrm{N\,m^{-2}}, \tag{B.4}$$

$$A_{\mathrm{n}}=\frac{m_{\mathrm{n}}^4c^5}{\hbar^3}=\frac{m_{\mathrm{n}}c^2}{\lambda_{\mathrm{n}}^3}=1.63\times10^{37}\ \mathrm{N\,m^{-2}}, \tag{B.5}$$

E. F. Keane, *The Transient Radio Sky*, Springer Theses,
DOI: 10.1007/978-3-642-19627-0,

$$B_{\rm e} = \frac{8\pi m_{\rm u} m_{\rm e}^3 c^3 \mu_{\rm e}}{3h^3} = (\mu_{\rm e} m_{\rm u}) \frac{1}{3\pi^2 \lambda_{\rm e}^3} = 1.9 \times 10^9 \text{ kg m}^{-3}, \tag{B.6}$$

$$B_{\rm n} = \frac{8\pi m_{\rm n}^4 c^3}{3h^3} = (m_{\rm n}) \frac{1}{3\pi^2 \lambda_{\rm n}^3} = 6.1 \times 10^{18} \text{ kg m}^{-3}, \tag{B.7}$$

B.3 Deutsch Fields

In Sect. 2.3.1 we derived the $\mathbf{E} \cdot \mathbf{B} = 0$ surfaces for the aligned case, $\alpha = 0$. The expressions for the fields in the general case, for the vacuum scenario, are known as the 'Deutsch fields' [3–5], and are given by:

$$B_r = 2B_0 \frac{a^3}{r^3} (\cos\alpha \cos\theta + \sin\alpha \sin\theta \cos\phi_s), \tag{B.8}$$

$$B_\theta = B_0 \frac{a^3}{r^3} (\cos\alpha \sin\theta - \sin\alpha \cos\theta \cos\phi_s), \tag{B.9}$$

$$B_\phi = B_0 \frac{a^3}{r^3} \sin\alpha \sin\phi_s, \tag{B.10}$$

$$E_r = \Omega B_0 a \left(\frac{2}{3} \cos\alpha \frac{a^2}{r^2} - 2\cos\alpha \frac{a^4}{r^4} P_2(\cos\theta) - 3\sin\alpha \frac{a^4}{r^4} \sin\theta \cos\theta \cos\phi_s \right), \tag{B.11}$$

$$E_\theta = \Omega B_0 a \left(-2\cos\alpha \frac{a^4}{r^4} \sin\theta \cos\theta + \sin\alpha \left(\frac{a^4}{r^4} \cos 2\theta - \frac{a^2}{r^2} \cos\phi_s \right) \right), \tag{B.12}$$

$$E_\phi = \Omega B_0 a \sin\alpha \left(\frac{a^2}{r^2} - \frac{a^4}{r^4} \right) \cos\theta \sin\phi_s, \tag{B.13}$$

where $\phi_s = \phi - \Omega t$. The expressions simplify somewhat if we examine in the frame co-rotating with the star, i.e. where $\phi_s = 0$. Then we can calculate $\mathbf{E} \cdot \mathbf{B} = 0$ surfaces for any choice of α. In Fig. 2.4, we showed the $\alpha = 0°$, $45°$ and $90°$ cases. We omit the algebra (which is trivial but lengthy) and simply state the results for the last two chosen cases.

$$\mathbf{E} \cdot \mathbf{B} = 2\Omega (B_0)^2 \frac{a^6}{r^5} \left[(\sin\theta + \cos\theta) \left(\frac{4}{3} + 2\frac{a^2}{r^2} (1 - 3\cos^2\theta - 3\cos\theta \sin\theta) \right) + (\cos\theta - \sin\theta) \left(1 + \frac{a^2}{r^2} (\sin 2\theta - \cos 2\theta) \right) \right] \quad (\alpha = 45°), \tag{B.14}$$

$$\mathbf{E} \cdot \mathbf{B} = 2\Omega (B_0)^2 \frac{a^6}{r^5} \cos\theta \left(1 - \frac{a^2}{r^2} (4\sin^2\theta + 1) \right) \quad (\alpha = 90°). \tag{B.15}$$

Appendix C
Birthrates: Supplementary

We quickly recount some recent work relevant to the discussion of neutron star evolution and birthrates. Ridley and Lorimer [13] have independently verified the results of Faucher-Giguère and Kaspi [9]. They have also included the alternative evolutionary model of Contopoulos and Spitkovsky [8], but in fact in this case, they found a lower figure of merit for the statistical reproduction of the known pulsars. They suggest that a new spin-down law is needed which produces random inclination angles, although this seems to be at odds with the observations of Weltevrede and Johnston [16]. Assuming an alignment timescale of $\sim 10^7$ year produced worse results. Another difficulty in modelling the evolution is the unknown functional form for the radio luminosity of pulsars. Faucher-Giguère and Kaspi [9] determine $L \propto P^{-1.5}\dot{P}^{0.5} \propto \dot{E}^{0.5}$, whereas Ridley and Lorimer [13] determine $L \propto P^{-1.0}\dot{P}^{0.5}$ as their best model, although there is certainly no striking trend on the $P - \dot{P}$ diagram and a large scatter from the best value would be needed (D. R. Lorimer, private communication). Soon to be published braking index measurements continue the trend of $1 \lesssim n \lesssim 3$ (C. M. Espinoza, private communication) but the simulations are insensitive to braking indices [13].

We can re-evaluate Eq. 3.11 which estimates the evolution time for a RRAT to evolve to become an XDINS, given our increased number of known sources. We find that $\langle P_{\rm RRAT} \rangle = 3.6\,{\rm s}$, identical to what was stated in Chap. 3, despite the doubling of known periods. $\langle P_{\rm XDINS} \rangle = 8.1\,{\rm s}$, unchanged as no new XDINS sources have been identified recently. The average RRAT period derivative has now changed quite a bit, with 14 measurements as opposed to just 3 for the initial estimate. It is also unclear as to what rate should be used to calculate an evolutionary time. What seems appropriate is to use the individual $\dot{P}$ values for each RRAT, i.e. if each RRAT i evolves into a 'typical' XDINS then the evolution time would be $t_{\rm i} = (\langle P_{\rm XDINS} \rangle - P_{\rm RRAT,i})(\langle \dot{P}_{\rm RRAT,i} \rangle)^{-1}$. For J1913+1330 (in the 'normal' region) this gives 26 Myr, for J1840−1419 (in the 'dying' region) it is 7.7 Myr and for J1854+0306 (in the 'high-B' region) it is 0.8 Myr. However it is not clear that all RRATs evolve to XDINSs (see e.g. the discussion of glitches in J1819−1458

E. F. Keane, *The Transient Radio Sky*, Springer Theses,
DOI: 10.1007/978-3-642-19627-0, © Springer-Verlag Berlin Heidelberg 2011

in Sect. 5.2.2), and the effects $\ddot{P}$ (possibly very large for dying pulsars in the Contopoulos and Spitkovsky [8] model) may need to be considered also.

A recent study by Schwab et al. [14] of the 14 best measured neutron star masses, a subset of those shown in Fig. 2.1, has shown that several were probably formed in electron capture supernova, although all in binary systems [14].

Recently the first measurement of a period derivative for a CCO has been performed. PSR J1852+0040, associated with the SNR Kestevan 79, has a measured period derivative of $\dot{P} = (8.68 \pm 0.09) \times 10^{-18}$ [10] implying, in the dipolar magnetic field scenario, the lowest magnetic field strength of any young neutron star of just $B_S = 3.1 \times 10^{10}$ G. This magnetic field strength is so low it has been suggested that it corresponds to the fossil field (i.e. from the supernova) but that, due to their slow birth periods, no dynamo action has occurred to increase the field strength [8, 15].

As mentioned later in the thesis, a radio pulsar with extreme flux modulation, PSR J1622−4950, with a magnetic field strength of $B \sim 3 \times 10^{14}$ G (the highest of all known radio pulsars), has recently been discovered [12]. It is thought to be a magnetar—the first discovered via radio emission.

As of late 2010, there are now four reliable XDINS $\dot{P}$ measurements for XDINSs and the extra corresponding characteristic ages are $\approx$3.6 and $\approx$3.7 and Myr respectively [11].

Appendix D
PMSingle: Supplementary

D.1 Single Pulse Searches

If the pulses from a pulsar have a flux density distribution $f(S)$, then $f(S)dS$ is the probability of there being a pulse in the flux density range between S and $S + dS$. The cumulative distribution $F(S)$, represents the probability of there being a pulse between 0 and S, so that $1 - F(S)$ is the probability of pulse, with flux density higher than S. If the probability of detecting a pulse of flux density S_{peak} or greater is $1/N$, then if we observe for N periods, the expected number of pulses with $S \geq S_{\mathrm{peak}}$ is 1, i.e. $1 - F(S_{\mathrm{peak}}) = N^{-1}$, or $F(S_{\mathrm{peak}}) = (N - 1)/N$, as stated in Eq. 4.5.

If S_{ave} is the mean peak intensity *which we measure* then it is given by:

$$S_{\mathrm{ave}} = \frac{\int_0^{S_{\mathrm{peak}}} Sf(S)dS}{\int_0^{S_{\mathrm{peak}}} f(S)dS} = \frac{\int_0^{S_{\mathrm{peak}}} Sf(S)dS}{F(S_{\mathrm{peak}})}, \tag{D.1}$$

$$= \left(\frac{N}{N-1}\right) \int\limits_0^{S_{\mathrm{peak}}} Sf(S)dS, \tag{D.2}$$

where we note that this is not the average of the distribution $f(S)$, as we do not have pulses with $S \to \infty$ in a finite observing time. The average (over period) flux density that we measure is then $\langle S \rangle = \zeta(W_{\mathrm{folded}}/P)$, where W_{folded} is the folded width, which in general is wider than the single pulse width $W_{\mathrm{folded}} = \varphi W$, and ζ is once again a shape-dependent constant (as pulses are not rectangular in general).

The FFT signal-to-noise ratio is given by $(S/N)_{\mathrm{FFT}} = \sqrt{n_{\mathrm{p}} BT} \langle S \rangle S_{\mathrm{sys}}^{-1} h_{\Sigma}$, where h_{Σ} is a factor due to harmonic summing (see e.g. Lorimer and Kramer [1]), which can be approximated as $h_{\Sigma} \approx (1/2)(P/W)^{1/2}$ [18]. The FFT S/N is then given by:

E. F. Keane, *The Transient Radio Sky*, Springer Theses,
DOI: 10.1007/978-3-642-19627-0, © Springer-Verlag Berlin Heidelberg 2011

$$(\mathrm{S/N})_{\mathrm{FFT}} = \frac{\zeta\varphi}{2}\sqrt{n_{\mathrm{p}}BW}S_{\mathrm{ave}}S_{\mathrm{sys}}^{-1}N^{1/2}. \tag{D.3}$$

Comparing this to the expression for the single pulse search S/N:

$$(\mathrm{S/N})_{\mathrm{peak}} = \eta\sqrt{n_{\mathrm{p}}BW}S_{\mathrm{peak}}S_{\mathrm{sys}}^{-1}, \tag{D.4}$$

we can see that their ratio, which we define as r is given by:

$$r = \frac{A}{N^{1/2}}\frac{S_{\mathrm{peak}}}{S_{\mathrm{ave}}}, \tag{D.5}$$

where $A = (2\eta)/(\zeta\varphi) \approx 2$ is a product of constants. Using this expression, we can simply determine S_{peak} and S_{ave} for any distribution $f(S)$ and we know $r = r(N)$, as we have done for the results listed in Sect. 4.2.

McLaughlin and Cordes [18] have performed these same calculations, except for the the lognormal case. We agree with their results in most cases, except for a few typos. Their Eq. A.13 differs from ours by a factor $(N/N-1)$, which is inconsequential and perhaps due to slightly differing definitions for the exponential distribution. Their Eq. A.18, the expression for the maximum flux density for a power law of the form $\alpha = 2$, has a power to which the expression is raised, which is in error by a factor of -1. This seems to be simply a typo, as a correct result for r is given subsequently.

Appendix E
Timing: Supplementary

E.1 Timing Standards

There are a number of timing systems which we need to be aware of when performing pulsar timing [19, 20]. *Terrestrial Time* (TT) is an ideal time system, the proper time on the geoid of Earth. *International Atomic Time* (TAI) is the principal realisation of TT and these are related by:

$$\mathrm{TT} = \mathrm{TAI} + 32.184\,\mathrm{s}, \tag{E.1}$$

where the offset is due to historical reasons. *Universal Coordinated Time* (UTC) is the time recorded by observatory clocks, and that which humans live their lives to, the replacement for *Greenwich Mean Time* (GMT) in 1972. TAI is defined as the average a large number of atomic clocks maintained by the Bureau International des Poids et Mesures (BIPM[1]). It is based on TAI (i.e. it has the same rate) but is updated with leap seconds to account for the slowing rotation of the Earth. TAI and UTC are related by:

$$\mathrm{TAI} = \mathrm{UTC} + 10\,\mathrm{s} + N_{\mathrm{LS}}, \tag{E.2}$$

where the 10 s offset is for historical reasons and N_{LS} is the number of leap seconds introduced since the beginning of 1972. At the time of writing, $N_{\mathrm{LS}} = 24\,\mathrm{s}$. When we record SATs at telescopes during pulsar observations, we do so in UTC. We must convert this first to TT. Next we convert it to *Barycentred Dynamical Time* (TDB), which varies from TT only by periodic 0.5-year and 1-year "Einstein delay" terms, which account for the changing atomic clock tick rates as the Earth moves through the gravitational potential of the solar system, like:

$$TDB = TT + (0.001658 * \sin\gamma + 0.000014 * \sin\gamma + \cdots)\,\mathrm{s}, \tag{E.3}$$

[1] http://www.bipm.org

E. F. Keane, *The Transient Radio Sky*, Springer Theses,
DOI: 10.1007/978-3-642-19627-0, © Springer-Verlag Berlin Heidelberg 2011

where $\gamma = 357.53° + 0.9856003°(JD - 2451545.0)$. We do not simply convert our SATs to TDB, in fact we convert them to TDB times at the solar system barycentre. We also remove the (variable) propagation delay due to the solar system gravitational potential and convert the times to 'infinite frequency', i.e. remove the dispersion delay.

$$T_{\mathrm{BAT,SSBC}}(TDB) = T_{\mathrm{SAT}}(TDB) + T_{\mathrm{R}} + T_{\mathrm{S}} + T_{\mathrm{E}} - T_{\mathrm{DM}}, \tag{E.4}$$

where T_{R} is the Röemer (geometrical) delay, T_{S} is the Shapiro (space–time curvature) delay and T_{E} is the Einstein delay. Recently, it has become common to convert TOAs to *Barycentred Coordinated Time* (TCB), which is a time system where the effects of the solar system gravitational potential have been removed. TCB and TDB are diverging at a constant rate since they were defined as coincident at a given instant in 1977. What this means is that older software, like PSRTIME and TEMPO use TDB, whereas the newest software, like TEMPO2, use TCB (or TDB). In practise, all the final timing results in this thesis are presented in TCB.

E.2 Pulse Profile Stability

If we consider a pulse profile $P(t)$ to consist of a noise-free template $T(t)$ with additive noise then we have $P(t) = T(t + \psi) + N(t)$, where we take the profile and templates to be appropriately scaled. If we perform the correlation coefficient R the templates must be aligned, i.e. $\psi = 0$ and R is defined as:

$$R = \frac{\mathrm{cov}(P, T)}{\sigma_{\mathrm{P}}\sigma_{\mathrm{T}}}, \tag{E.5}$$

for profile and template with M bins.

$$R = \frac{\sum_{i=0}^{M-1}(P_{\mathrm{i}} - \bar{P})(T_{\mathrm{i}} - \bar{T})}{\sqrt{\sum_{i=0}^{M-1}(P_{\mathrm{i}} - \bar{P})\sum_{i=0}^{M-1}(T_{\mathrm{i}} - \bar{T})}}. \tag{E.6}$$

We let $X_{\mathrm{i}} = P_{\mathrm{i}} - \bar{P} = T_{\mathrm{i}} - \bar{T} + N_{\mathrm{i}} - \bar{N}$, and we can define $\bar{N} = 0$ without loss of generality, so that this becomes:

$$R = \frac{\sum_{i=0}^{M-1} X_{\mathrm{i}}^2 + X_{\mathrm{i}}N_{\mathrm{i}}}{\sqrt{\sum_{i=0}^{M-1} X_{\mathrm{i}}^2 \sum_{i=0}^{M-1}(X_{\mathrm{i}}^2 - 2X_{\mathrm{i}}N_{\mathrm{i}} + N_{\mathrm{i}}^2)}}. \tag{E.7}$$

The $X_{\mathrm{i}}N_{\mathrm{i}}$ cross-terms vanish when we go to the expectation value so that we have:

$$\langle R \rangle = \frac{\sum_{i=0}^{M-1} X_{\mathrm{i}}^2}{\sqrt{\sum_{i=0}^{M-1} X_{\mathrm{i}}^2 \sum_{i=0}^{M-1}(X_{\mathrm{i}}^2 + N_{\mathrm{i}}^2)}} = \left(1 + \frac{\sum_{i=0}^{M-1} N_{\mathrm{i}}^2}{\sum_{i=0}^{M-1} X_{\mathrm{i}}^2}\right)^{-\frac{1}{2}}. \tag{E.8}$$

Expanding in the second term which is $\ll 1$ we get:

$$\langle 1 - R \rangle = 1 - \left(1 - \frac{K}{2}\frac{\sum_{i=0}^{M-1} N_{\rm i}^2}{M} + \cdots\right) = \frac{K}{2}(\text{rms})^2. \tag{E.9}$$

The rms of the noise depends on the signal-to-noise ratio and hence on the number of pulse periods, n, added to create $P(t)$ so that this is equivalent to:

$$\langle 1 - R \rangle \propto K n^{-1}. \tag{E.10}$$

Of course the $K = M/\sum_{i=0}^{M-1} X_{\rm i}^2$ term itself has some dependence on noise (and therefore n) so that actually we get behaviour of the type $a/(a+n)$. However the constant a is typically much smaller than n (i.e. the rms of the noise-free template is much higher than the rms of the noise for high signal-to-noise values) so that we get $\langle 1 - R(n) \rangle \propto n^{-1}$ for all cases of interest for pulsar profiles.

References

1. D.R. Lorimer, M. Kramer, *Handbook of Pulsar Astronomy* (Cambridge University Press, Cambridge, 2005)
2. R.A. Osten, T.S. Bastian, ApJ **674**, 1078 (2008)
3. A.J. Deutsch, Ann. d'Astrophys. **18**, 1 (1955)
4. J. McDonald, A. Shearer, ApJ **690**, 13 (2009)
5. F.C. Michel, H. Li, Phys. Rep. **318**, 227 (1999)
6. C.W. Misner, K.S. Thorne, J.A. Wheeler, Gravitation (W. H. Freeman, San Francisco, 1973)
7. S.L. Shapiro, S.A. Teukolsky, *Black Holes, White Dwarfs and Neutron Stars*. The Physics of Compact Objects (Wiley-Interscience, New York, 1983)
8. I. Contopoulos, A. Spitkovsky, ApJ **643**, 1139 (2006)
9. C.A. Faucher-Giguère, V.M. Kaspi, ApJ **643**, 332 (2006)
10. J.P. Halpern, E.V. Gotthelf, ApJ **709**, 436 (2010)
11. D.L. Kaplan, van M.H. Kerkwijk, ApJ **705**, 798 (2009)
12. L. Levin et al., ArXiv e-prints (astro-ph/1007.1052) (2010)
13. J.P. Ridley, D.R. Lorimer, MNRAS **404**, 1081 (2010)
14. J. Schwab, P. Podsiadlowski, S. Rappaport, ApJ **719**, 722 (2010)
15. C. Thompson, R.C. Duncan, ApJ **408**, 194 (1993)
16. P. Weltevrede, S. Johnston, MNRAS **387**, 1755 (2008)
17. D.R. Lorimer, M. Kramer, *Handbook of Pulsar Astronomy* (Cambridge University Press, Cambridge, 2005)
18. M.A. McLaughlin, J.M. Cordes, ApJ **596**, 982 (2003)
19. R.T. Edwards, G.B. Hobbs, R.N. Manchester, MNRAS **372**, 1549 (2006)
20. G.B. Hobbs, R.T. Edwards, R.N. Manchester, MNRAS **369**, 655 (2006)

The Author

The author graduated from the National University of Ireland, Galway in 2006 with a First class B.Sc. Hons degree in Physics and Astronomy before receiving a Master in Advanced Study of Mathematics (aka "Part 3") with merit from Trinity College, University of Cambridge, in 2007. The work presented in this thesis has been undertaken at the Jodrell Bank Centre for Astrophysics, in The University of Manchester, during a period of three years. This work has resulted in the following publications, as well as conference contributions in 12 different countries (20 oral presentations and 10 poster presentations).

Refereed Publications

- "On the birthrates of Galactic neutron stars", **E. F. Keane** and M. Kramer, 2008, MNRAS, 391, pp. 2009–2016 (astro-ph/0810.1512).
- "An Interference Removal Technique for Radio Pulsar Searches", R. P. Eatough, **E. F. Keane** and A. G. Lyne, 2009, MNRAS, 395, pp. 410–415 (astro-ph/0901.3993).
- "Timing Observations of Rotating Radio Transients", M. A. McLaughlin, A. G. Lyne, **E. F. Keane**, M. Kramer, J. J. Miller, D. R. Lorimer and R. N. Manchester, 2009, MNRAS, 400, pp. 1431–1438 (astro-ph/0908.3813).
- "Unusual Glitch Activity in the RRAT J1819–1458", A. G. Lyne, M. A. McLaughlin, **E. F. Keane**, M. Kramer, C. M. Espinoza, B. W. Stappers, N. T. Palliyaguru and J. J. Miller, 2009, MNRAS, 400, pp. 1439–1444 (astro-ph/ 0909.1165).
- "Further Searches for RRATs in the Parkes Multi-Beam Pulsar Survey", **E. F. Keane**, D. A. Ludovici, R. P. Eatough, M. Kramer, A. G. Lyne, M. A. McLaughlin and B. W. Stappers, 2009, MNRAS, 401, pp. 1057–1068 (astro-ph/0909.1924).

DOI: 10.1007/978-3-642-19627-0,

- "The European Pulsar Timing Array: current efforts and a LEAP toward the future", R. D. Ferdman et al. (1 of 25 authors), 2010, CQG, 27, pp. 084014 (astro-ph/1003.3405).

Conference Proceedings

- "Transient Radio Neutron Stars", **E. F. Keane**, 2010, to be published in the proceedings of "High Time Resolution Astrophysics IV—The Era of Extremely Large Telescopes", held on May 5–7, 2010, Agios Nikolaos, Greece (astro-ph/1008.3693).

In Preparation

- "Timing Solutions of Rotating Radio Transients", **E. F. Keane**, M. Kramer, A. G. Lyne, B. W. Stappers and M. A. McLaughlin, MNRAS, submitted.
- "Radio Properties of Rotating Radio Transients I: Spectral Indices and Pulse Amplitude Distributions", J. J. Miller, M. A. McLaughlin, **E. F. Keane**, A. G. Lyne, M. Kramer, D. R. Lorimer, R. N. Manchester, F. Camilo and I. H. Stairs, MNRAS, submitted.
- "A search for optical bursts from RRAT J1819–1458: II. Simultaneous ULTRACAM-Lovell Telescope observations", V. S. Dhillon, **E. F. Keane**, T. R. Marsh, B. W. Stappers, C. M. Copperwheat, R. D. G. Hickman, C. A. Jordan, P. Kerry, M. Kramer, S. P. Littlefair, A. G. Lyne, R. P. Mignani, A. Shearer, MNRAS, submitted.
- "Radio Properties of Rotating Radio Transients II: Searches for Randomness and Periodicities in Pulse Arrival Times", N. T. Palliyaguru, M. A. McLaughlin, **E. F. Keane**, M. Kramer, A. G. Lyne, D. R. Lorimer, R. N. Manchester, F. Camilo and I. H. Stairs, MNRAS, submitted.

Index

A
Accretion induced collapse, 57
Amplitude distribution, 63
  bimodal/nulling, 66
  exponential, 64, 65
  lognormal, 64, 65
  power law, 65, 66
AXPs, *see* Magnetars

B
Birthrates
  birthrate problem, 50, 51, 52, 53, 54, 55, 56, 57
  CCOs, 49
  magnetars, 49, 50
  pulsar current analysis, 46, 48
  radio pulsars, 46, 48
  RRATs, 48
  XDINSs, 49, 50
Brightness temperature, 1

C
CCOs, 45, 180
  birthrates, 49
CCSN, *see* Core-collapse supernova rate
Core-collapse supernova rate, 45

D
DFB, 116, 117
  bandpass, 118
  clee hill radar station, 117, 121
  RRAT observations, 117
Dispersion measure, 70

E
Electron capture supernova, 56, 180

F
Flux density distribution, *see* Amplitude distribution

G
Glitches, 109
  changes in $\dot{\nu}$, 114
  J1819–1458, 109
  vortex unpinning, 24

H
HYDRA
  specifications, 69

I
IMF, *see* Initial mass function
Initial mass function, 16, 17, 18
  kroupa IMF, 16

J
J0735–62
  confirmation, 130
J0845–36
  discovery, 84, 87
J0847–4316
  timing solution, 124
  X-ray limit, 151
J1047–58
  discovery, 87, 84

E. F. Keane, *The Transient Radio Sky*, Springer Theses,
DOI: 10.1007/978-3-642-19627-0, © Springer-Verlag Berlin Heidelberg 2011

J (*cont.*)
J1111–55
  discovery, 84, 87
J1226–32
  confirmation, 130
J1308–67
  discovery, 84, 87
  dispersion sweep, 92
J1311–59
  discovery, 84, 87
  dispersion sweep, 92
J1317–5759
  infrared limit, 152
  timing solution, 124
  X-ray limit, 151
J1404–58
  discovery, 87, 84
J1423–56
  discovery, 87, 84
J1444–6026
  timing solution, 124
J1513–5946
  discovery, 87, 84, 85
  timing residuals, 127
  timing solution, 123, 124
J1554–5209
  discovery, 87, 84, 85
  timing residuals, 127
  timing solution, 123, 126, 124
J1649–46
  discovery, 87, 84
  dispersion sweep, 92
J1652–4406
  discovery, 87, 84
  timing solution, 126, 124
J1654–23
  confirmation, 131
J1703–38
  discovery, 87, 84, 85, 89
J1707–4417
  discovery, 87, 84, 85
  timing residuals, 127
  timing solution, 126, 127, 124
J1724–35
  discovery, 87, 84, 89
J1727–29
  discovery, 87, 84, 89
J1807–2557
  discovery, 87, 84, 89
  timing residuals, 127
  timing solution, 127, 128, 124
J1819–1458
  exhausted magnetar?, 112, 113, 114
  fermi detection, 114
  glitches, 109, 110, 112, 113, 114
  infrared observations, 152
  jodrell bank observations, 109, 110, 113, 114
  optical observations, 140, 142, 143, 145, 146
  polarisation properties, 150
  timing residuals, 110
  timing solution, 124
  X-ray observations, 33, 151, 152
J1826–1419
  timing solution, 124
J1840–1419
  chandra observations, 137, 139
  discovery, 87
  timing residuals, 127
  timing solution, 128, 124
J1846–0257
  binary scenario, 152
  timing solution, 124
  X-ray limit, 151
J1852–08
  discovery, 87, 84
  dispersion sweep, 92
  origin, 92, 91
J1854+0306
  discovery, 87, 84, 92
  PALFA, 150
  timing residuals, 167, 127
  timing solution, 128, 124
J1911+00
  X-ray candidate, 33
J1913+1330
  jodrell bank observations, 115
  timing residuals, 116
  timing solution, 116, 124

L
Life and death of stars, 15

M
Magnetars, 45
  birthrates, 49, 50
  field estimates, 154
  radio, 9, 45, 150, 180
Millisecond pulsars, 42
MSPs, *see* Millisecond pulsars

N
Neutron stars
  back of an envelope, 18, 19, 20, 21, 23, 25, 26, 27, 32, 33, 36, 37
  core composition, 25
  Coulomb lattice, 22
  internal structure, 19, 175
  inverse β decay, 22
  minimum radius, 175
  neutron drip line, 23
  neutronisation, 23
  superfluid vortices, 23, 24
  vortex pinning, 24
New technology telescope, 140, 146, 142, 143, 144

P
Parkes multi-beam pulsar survey, 67, 69, 74, 71, 72, 73, 74, 76, 77, 78, 79, 82, 85, 87, 88, 89, 91, 93, 98
  multi-beam receiver, 67
  survey parameters, 69
Period-period derivative diagram
  2010, top half, 138
  2008
  2009
  2010
PMPS, *see* Parkes multi-beam pulsar survey
PMSingle, 76, 77, 84, 85, 87, 96, 97, 98
  amplitude distributions, 94, 95, 96, 97
  new discoveries, 80, 81, 87, 90
  processing steps, 77
  timing solutions, 123, 128, 124, 127, 129, 130, 132, 133, 135
Pulsar timing, *see* Timing
Pulsars
  birthrates, 46, 47
  braking index, 28, 135, 179
  death line, 135, 137
  death valley, 153
  deutsch fields, 176
  force-free model, 31, 32
  intermittent, 153
  light cylinder, 30
  lighthouse model, 27
  magnetospheric switching, 153
  moding, 101
  nulling, 101, 153, 162, 163
  photon splitting, 154
  pulsar equation, 31
  sub-pulse drifting, 101
  switching magnetospheres, 162, 163
  vacuum model, 28, 30, 31
  when a pulsar is a RRAT, 158

R
Radio transients
  auroral radio emission, 7
  known sources, 6, 10
  Lorimer burst, 8
  phase space, 4, 5
  theoretical sources, 10
RFI
  clee hill radar station, 117, 121
  directionality at jodrell bank, 120
  impulsive, 70
  jodrell bank, 120
RRATs, 43
  2007 status, 32, 33, 37
  2010 status, 132
  answers, 164, 165, 166, 167
  birthraes, 47, 48
  definition, 158
  facts, 167
  misconceptions, 167
  non-PMPS discoveries, 150
  questions, 37
  radio spectra, 150
  special, 168
  timing, 104, 103, 106, 107

S
SGRs, *see* Magnetars
Single pulse searches, 62, 63, 64, 65, 66, 181, 182

T
Thesis
  overview, 157
  questions for the future, 168
Timing, 101, 102, 103, 104, 106, 107
  assumption, 102
  PSRCHIVE, 118
  PSRPROF, 105
  PSRTIME, 106, 184
  pulsars, 102
  pulse profile stability, 102, 184, 185
  single pulse timing, 104, 105, 125, 127, 128, 130, 132
  standards
    TAI, 183
    TCB, 106, 184

T (*cont.*)
TDB, 183, 106
TT, 183
UTC, 183, 106
TEMPO, 184
TEMPO2, 184
the perils of EFAC, 131, 132

U
ULTRACAM, 140, 141, 143, 145, 146

V
Virial theorem, 15

W
William Herschel telescope, 140, 141, 142, 143, 146, 147

X
X-ray binaries, 42
XDINSs, 44, 180
birthrates, 49

Z
Zero-DM filter, 70, 71, 73, 74, 75
examples, 74
Lorimer burst, 74
response function